RAMPAGE

ALSO BY PETER BRIGGS

Mysteries of Our World
Buccaneer Harbor
Rivers in the Sea
The Great Global Rift
Science Ship
Men in the Sea
Water: The Vital Essence
Laboratory at the Bottom of the World
200 Million Years Beneath the Sea
Will California Fall into the Sea?
What is the Grand Design?

THE STORY OF
DISASTROUS FLOODS,
BROKEN DAMS, AND
HUMAN FALLIBILITY

RAMPAGE

PETER
BRIGGS

DAVID McKAY COMPANY, INC.
NEW YORK

RAMPAGE

Copyright © 1973 by Peter Briggs

LIBRARY OF CONGRESS CATALOG CARD NUMBER: 73–76557
manufactured in the United States of America
Designed by E. O'Connor

Contents

Introduction

"Can we help you?" the woman asked.

"Tell me what I am going to do now," answered a man who has lost his wife and two of his four children and all his earthly possessions. Craig Ammerman, Logan (West Virginia) *Banner,* Feb. 28, 1972

Since Hernando de Soto discovered the Mississippi River in 1543, and was prevented from crossing it for a month because of an enormous spring flood, great excess flows of water have always been a problem in what is now the United States. The problem grows more serious all the time, in spite of the fact that the Federal government with some local help has spent about $30 billion on flood control since the

Army Engineers began trying to tame the Mississippi in the last century.

From the time records began to be kept, the average number of floods has been about the same and in no recent flood has there been anything like the 2,100 lives lost in the Johnstown, Pennsylvania flood of 1889; though the individual grief has not diminished for those who lose family or friends today. The great difference today is that floods are so much more costly. Due to inflation, partly, but also because more and more homes, more industry, more highways and bridges have been built on the extremely dangerous flood plains. Whoever builds on the river's traditional province risks its mindless vengeance when the flow is unusually high.

A very similar risk is taken by everyone who builds a summer cottage or a resort facility anywhere along the thousands of miles of United States coastline. Hurricanes can be expected in the Atlantic and the Gulf of Mexico anywhere from Masschusetts, past Florida to Texas, and then beyond to Mexico and Central America. By the definition of floods "a great flow of water . . . overflowing land" hurricanes with their abnormally high tides and huge waves certainly cause floods.

Another hazard exists for those who live or work below dams or reservoirs. Such structures need very sophisticated engineering but there are evidently thousands of dams and reservoirs that do not live up to minimum specifications of safety. A surprising number in the United States have given way because they were simply too weak to begin with, because the base has subsided or was placed on "incompetent" rocks, or because there was more rainfall or a stronger earthquake than planned in the specifications. In addition, dams have been neglected, even forgotten.

Under a fairly new law, presidents have been declaring many stricken regions as "disaster areas." This alarming term has meant, primarily, Federal provision for low-cost loans. Disasters come quite frequently in the United States. In the month of January, 1972, alone there were disasters declared in California, Oklahoma, Mississippi, Oregon, Texas, Arkansas and the State of Washington. First allocations of money amounted to about $12 million. All of these disasters were caused by severe storms and heavy rains which caused flooding. All this suggests that floods are a problem of some dimension but the United States is so big and, comparatively, so few people are affected by any flood that bringing the matter up may only seem alarmist.

However, the present Federal budget every year for flood prevention is at least $600 million. And, for the year 1972, in August the president asked Congress for *$1.8 billion for flood relief.* Such sums are, or should be (one supposes), of concern to the taxpayer. In addition, 14 million people in the United States live upon flood plains. No one has ever estimated how many live below reservoirs and dams. Or on coast lines exposed to hurricane floods. Perhaps one out of every ten persons could be hit a staggering blow by a flood from one source or another.

Thus, 20 million Americans are exposed to flood danger. Except in the case of a dam bursting, there will probably be no tremendous loss of life in all the many floods we can expect, but there will certainly be losses of property in the billions of dollars. The 180 million people who live in safety from floods are thus subsidizing those who risk their homes and factories to the destructive water. With large government subsidies, those who are wiped out will certainly return to the same places and those who are bailed out will some day need the help of our money again. Is this reasonable? Should not an area be zoned against construction once a flood has swept it clean? Yet who can deprive people of property to which they have title?

Even if it were constitutional to do so, it would be impossible physically and economically to move 20 million people away from danger. Flood plains are usually the best farm land, the most desirable zones for factories, for transportation, and often extremely attractive places to build houses.

These are the bare bones of the dilemma. Any solutions will only be compromises—but some measures to prevent tragedy can be taken. One of the measures is to eliminate the possibility of any more dam failures. Four in a period of sixteen months seems unpardonable.

RAMPAGE

What Happened on Buffalo Creek

The disaster year of 1972 began to hit full stride on February 26. Logan County in West Virginia, today in the heart of the strip coal mining territory, had had snowfall up in the mountains the previous weekend and then, beginning on Wednesday, three days of heavy rain. Buffalo Creek, a stream which gives its name to one of the many narrow hollows in the county, began to rise but most residents of the fifteen small communities on Buffalo Creek did not give it much thought. They all knew there was an artificial, filthy lake about three-quarters of a mile long, up above them towards the top of the hill. The lake was caused by a slag heap of mining wastes, piled into a dam 150 feet high, that had never been intended as a dam but had simply accumulated there. The waste pile had given way once in 1966 and caused a small flood, but officials of the Buffalo Mining Company, which operated the strip mine at the top of the hollow, assured them the dam was safe. One employee of the Buffalo company had often speculated about the heap giving way, but he enjoyed operating his power boat on the small lake. The dam was perhaps thirty years old but maybe older than that. No one was certain.

About 5 o'clock on Saturday morning, February 26, Otto Mutters, Deputy Sheriff in the Buffalo Creek town of Man, received a phone call saying that the slag pile was about to give way. Sheriff Mutters immediately got into his car and drove along the narrow, blacktop road towards the dam. He followed Steve Dasovich, superintendent of the mining company, up the road. They got to the foot of the mountain, at the base of the slag heap about 6:10 A.M. Superintendent Dasovich climbed up the heap alone and came down in a few minutes, saying a ditch had been dug around the heap and everything would now be all right.

Sheriff Mutters said, the following day, "It just seemed like it might possibly break to me." On his way down the hill, he began warning the residents. He stopped at the Trent household, but they paid no attention. "I hear they lost three sons, a mother, and a daughter that I know of." He continued giving the alarm all along the 14 miles of Buffalo Creek. Those who believed the sheriff's warning began to follow him down the road in their cars, blowing their horns and blinking lights to alert valley residents.

No one seems to know the exact moment when the untended dam broke, up in the hills, but at about 8 A.M. Kenny Vance, a neighbor of the Linville family in Amherstdale, ran over to tell the Linvilles that the slag pile had burst. Butch Linville remembered 1963 when Buffalo Creek only came into their yard. He did not worry for his parents, his wife or his daughter, who all shared the same house, but thought about his brother who lived right beside the creek near the town of Man. With his father, Butch Linville jumped into his car to hurry and warn Leon, the brother, Father and son only reached the railroad tracks when they saw a wall of water rushing down the valley. They scrambled out of the automobile as fast as they could and started up the hill. Halfway up, Butch Linville got down on his knees and prayed for his wife and baby and mother. Then he looked up and saw them all, safe on the hill on the other side of the valley. (Buffalo Creek valley is not usually more than 100 yards wide.)

Almost immediately after Butch and his father got safely out of the car, they saw the water pick it up and slam the vehicle into a building, which collapsed over it. They saw the water carry their house 20 feet until it smashed into another house. "I just didn't believe I would see that much water and houses floating down the creek," Butch said. "Men and women, children and animals also were floating in that high wall of water. I saw this woman and baby floating in the water holding on to a tree limb. Someone got them

3

out and they were okay. I don't see how they lived through it."

Survivors said the wall of water just below the dam was 50 feet high. Further down it was only 15 to 25 feet high. No one has any idea how much impounded water was released nor how fast it flowed. The black, slimy dirt left by the water could later be seen on the second story of buildings ten miles downstream.

Bill Mays of Lundale told a local newspaperman, "I saw this large wall of water bringing logs and houses down through the community. When I saw it coming, I tried to warn some of my neighbors. All of them got out except one woman. She just looked at me. She must have been in shock." Mays found refuge in the Man High School. There he told how he saw the wall of water take his mobile home and wrap it around a bridge. "I don't have anywhere to go now. My wife is in the hospital, having a baby. We don't have a home and I don't know when I will get to go back to work." Mays was employed by the Buffalo Mining Company.

Mrs. Carolyn Harvey had enough warning to climb the hill with her two children, her husband, and some clothes, to reach her parents' house an hour before the water hit. When she came down the hill the following day, her mobile home was gone but she did find a salt and pepper shaker, a sugar bowl, and two pitchers worth salvaging. She also found a couple of pairs of her husband's pants which she thought might be cleaned and saved. And nothing more.

When Delbert Sparks returned to his 4.5 acre farm which had had two houses and two cars, there was nothing. "Just one big old pine tree. I didn't even have time to get my billfold out of my night table. It's all gone. I worked 40 years and it's all leveled and gone."

From an initial report of 24 deaths, the toll kept mounting day by day until it approached 100 persons with many never to be accounted for; buried beneath thousands of tons of slag and mud.

Relief measures by the Red Cross and the National Guard began the same day the flood occurred. Criticism of the mining company, the state of West Virginia, and the Federal government began the next day. The *New York Times* had a story from Charleston, West Virginia that next day headlined, "Blame for Flood Is Hard to Fix."

But, before the carping, the generous actions that are almost always peoples' initial response to disaster. The young men of the National Guard were early on the scene, slogging through the black mud, on foot, by jeep and truck, looking for survivors and for bodies. No complaints were heard about this disasteful task. Since most of the roads were gone, the vehicles traveled up and down Buffalo Creek itself, now reduced to its normal shallow flow. Many of the bodies were flown out by helicopter and left in the parking lot of the hospital in the town of Man. The Guardsmen burned hundreds of piles of debris, much of it reeking from the smell of decayed flesh. Coal workers from the mining company operated huge machines, gently prodding the debris to help the Guardsmen search for corpses. The mining company also brought in generators and strung electric cables to those buildings still standing.

The West Virginia State Health Director declared Buffalo Creek, filled with mud, ruins, and the smell of death, a health and safety hazard. Typhoid shots were given to every survivor who could be corralled. (Warm temperatures three days after the flood brought out rattlesnakes and five people were bitten.)

Among the volunteers were carpenters sent from the Mennonite Disaster Service in Lancaster, Pennsylvania who worked (ultimately for months) on the enormous job of rebuilding. Seventh Day Adventists gave out food and clothing. Underwater recovery teams from the Sheriff's Office in Erie, Pennsylvania came to join in the grisly search for bodies. The Salvation Army set up 7 mobile canteens and 67 officers, from 9 states, with 263 volunteers served more than 200,000 meals. They also operated 4 clothing and food centers, aiding 5,706 families.

The Tridelphia Women's Club of West Virginia set up a Buffalo Creek Disaster Fund and solicited donations. Among the contributors were Weight Watchers, Inc., which sent a check for $1,000. It was decided that the first thing was to buy school supplies for students who lost everything in the flood. A District of Columbia group calling themselves the Department of Army Wives collected 80 boxes of children's clothing and, led by Mrs. Bruce Palmer, wife of the Army's Vice Chief of Staff, the gifts and some of the ladies were flown by three Army helicopters to the Man High School for delivery of the clothing. This generosity turned out to be something of a burden, what with sorting out, cleaning, and fitting the correct size to the right child. The Seventh Day Adventists set up three sites to give away cooking utensils and tableware but it was finally announced that what the victims really needed was cash. The State Highway Department had quickly built a temporary road into the valley and new goods could be brought in.

Criticism of the Red Cross developed, even though they spent more than $800,000 cash. Some of the money went in direct outlay to individuals, 1,336 of them. The rest went into operating four mass care centers for 1,500 people. The Red Cross fed more than 12,000 disaster victims, many recovery workers, and answered thousands of inquiries, many of them from Vietnam. The Red Cross also provided long-term counseling about rehabilitation.

Don Hannah of the Red Cross explained why there were complaints. The organization tries to keep as complete records as possible since they feel obligated to make an accounting to their contributors, all of whom are voluntary. Thus, they ask questions about eligibility and need. "Some people resent this questioning," Hannah said, "and in some cases, the person being questioned will leave after the first question without giving us a chance to explain our procedure."

In some cases, those asking for help admitted they were not affected by the flooding but would say, "You gave help to my neighbor and he wasn't affected, either." Hannah believed that 25 percent of the people who received help did not need or deserve it.

"You might also hear complaints that some of the used clothing sent into the area was sent to other areas by the Red Cross. If you do hear that, it is correct." It is cheaper to buy new clothing than go through all the labor of working with the used clothing. In addition, giving away used clothing eliminates sales by local businesses, which also have a recovery problem.

The Federal government joined in the recovery effort at Buffalo Creek with a much better coordinated program than in other recent disasters, such as the San Fernando earthquake of 1971. A Disaster Declaration was ordered almost immediately and the Office of Emergency Preparedness made an initial allocation of $2,400,000. When the true extent of the damage was later realized, this amount was enormously increased. One coordinating officer was appointed to oversee all stages of the U.S. recovery effort.

The Army Corps of Engineers removed 300,000 cubic yards of debris to three sanitary landfill sites, then graded the sites and planted them with seed. This cost about $2 million. The Corps also prepared, at West Virginia's expense, 600 sites for mobile homes.

They, in addition, built eight temporary bridges in Buffalo Creek valley. The Small Business Administration, usually the slowest branch of the Government in previous disasters, had approved 504 home loans out of 581 applications by the middle of May. The loans came to a bit more than $2 million.

The Department of Health, Education, and Welfare expended $600 thousand for repairs to 15 schools. The Department of Labor found 531 jobs for 655 applicants and dispersed unemployment checks to 1,046 people. The Department of Transportation did emergency repair work on 18 miles of road. Agriculture provided food stamps for 6,500 people. It also let out a contract to seed all affected areas at a cost of $117,000. HUD produced 539 mobile homes by the middle of May, enough to house every family needing one. These mobile homes in past disasters have always been a problem. Usually there are not enough of them available immediately. People have to wait. In addition, the mobile homes present long-term financial problems. In this case, the decision was to house the occupants rent free, but for twelve months only. The Internal Revenue Service moved personnel into the valley to show residents how to file loss claims on their income tax and substantiate the losses. (The state and county also took measures to provide tax relief for the victims.)

Historically, as will be shown, the Federal government has been very slow to get into the disaster business. The military would be brought in to rescue people and establish martial law, against disorder, but after the first shock, people were left to fend for themselves or seek private charity. The American people have always been very generous. Millions were raised for relief of San Francisco after the earthquake and fire, but the moments of compassion were rather a Sunday thing and the Christian gesture forgotten by Monday.

With various laws passed by Congress in recent years, the executive department is obliged to take action and by the time of the Buffalo Creek flood of 1972 had learned to do it generously and efficiently.

All this goodness and light, however, did not produce a cordial atmosphere about the flood. The victims looked for a villain and he was easy to find.

Symptomatic of the obvious was Steve Dasovich, Buffalo Mining's vice president of operations, who had climbed up to the dam at 6:10 the morning of the flood and come down to tell Deputy Sheriff Mutters that there was no danger of a collapse. Mr. Dasovich went into shock when the flood struck and was taken to the hospital in Man and put under heavy sedation.

Day by day, the reaction grew against Buffalo Mining and its parent, the Pittston Company of New York. Yet the legal responsibility seemed hard to fix. United States Geological Survey said that a lack of either Federal or state regulations concerning coal industry impoundments of water made the hazards of such dam failures common all through the Appalachian coal fields. Since 1966, there had been three others dam failures in West Virginia, similar to Buffalo Creek. In the valley, people said that the company had received warnings long before the break that the crude dam was endangered by the heavy rains.

Still Federal officials, acknowledging that someone had made a "tragic mistake," did not feel sure a law had been violated. A two-year-old Federal coal Mine Health and Safety Act said that a coal mine settling basin or other surface water impoundment considered a "hazard" shall "be inspected at least once a week" by officials of the company responsible for it. But, asked the Bureau of Mines, did anyone know until Saturday that it was a hazard?

Some people in the government certainly did. The day after the disaster, William Davis of the Geological Survey in Washington issued a statement recalling that as long ago as 1966 a government study located "at least 75" poorly constructed waste-pile dams across natural narrows at Appalachian coal mines in five states. These impoundments are used to provide fresh water for coal preparation plants at the mouth of the mine and to settle out waste matter in the wash water after it has been used. The study was made after a slate bank near a coal mine in Wales, after a heavy rain, came crashing down onto a school house, killing 116 children and 5 teachers. (The British discovered later that they had no laws concerning the safety of such waste piles.)

Of the 75 hazardous mines in the Appalachians, 25 were in Logan and Mingo counties in West Virginia.

Davis said, "Many of these dams were not engineered to serve as dams and they generally lacked spillways or overflow channels. Such waste banks are susceptible to breaching following a heavy rain of more than three inches in a day." The Weather Bureau said that Logan County had had 4.7 inches of rain in the four day period before the flood. In the original report the Buffalo Creek dam was declared "Unstable. . . . Could be overtopped and breached. Flood and debris would damage church and two or three houses downstream, cover road and wash out railroad. . . . Lake contains about five million cubic feet of water." This report was sent out in a three-page letter, signed by Secretary of the Interior Udall, in 1966 to numerous officials in states with coal mines. The governor of West Virginia received a copy of the letter but the present governor, Arch A. Moore, Jr., pointed out that he was not governor in 1966 and had never heard about the matter.

As funerals began for the first 71 known dead, Mayor Rayman Herman of the town of Man said, "I think 99 percent of the people want to go back. They don't give up very easy." The mayor said he believed the flood was "an act of God." and that Buffalo Mining should not be blamed because, "You can't say it's someone's fault. There should not be tougher regulations just to make it tougher on the coal companies. Some regulations in the past have been unfair to the coal companies because they were made by people not real knowledgeable about the coal mining business." Mayor Herman had a general store in the valley and a working arrangement with the Buffalo Mining Company by which they would make payroll deductions from employees to pay bills owed to Herman.

Two days after the flood a vice president of the mining company spoke of "the flood, which we believe, of course, to be an act of God." The general superintendent of Buffalo Mining explained the slate pile by saying that state law forbade mining companies from dumping refuse in streams. "They were afraid of killing the trout."

Then, amidst all the statements being made, the Charleston, West Virginia, *Gazette* unearthed a forgotten 1913 state law which prohibits the erection of "any obstruction" more than 15 feet high across "any watercourse" without the engineering, design, and location approval of the State Public Service Commission. Apparently no coal company ever paid any attention to this law. The dams were simply built according to the convenience of dump truck drivers.

During the week after the disaster, Governor Moore appointed a nine-man committee to investigate. In the group were the heads of the state departments of Mining, Natural Resources, and the Public Service Commission, all of whose actions were being questioned. Also on the committee was the editor of the Logan *Banner*, the only person in the group from the area affected. He promised there would be no "whitewash."

In other developments, a conservationist group in Charleston called Citizens Against Strip Mining said they had recently made a series of tape-recorded interviews in the Buffalo Creek hollow. Residents were said to have described the refuse dumps and strip mining as "a cocked pistol" ready at any moment to loose a torrent of flood waters.

Mrs. Irene Rose, whose husband worked for Buffalo, was quoted as saying, "We've only been here a few years but we never would have moved here if we'd known about that dump. They kept telling us lies about it. Now I think as soon as we can get our stuff out of our house, this is one girl who isn't going to stay here. If they can lie to you once, they can lie to you again."

Ralph Nader, a prominent crusader against corporations, had his say on the situation from Washington. He charged that 350,000 people in West Virginia might be threatened by flooding from unstable coal mine dams like the one in Buffalo Creek. He demanded a Congressional investigation of the mining industries' dereliction throughout Appalachia. He added "The Buffalo Creek massacre is only one more in the long series of tragedies which coal corporations have perpetrated." He said that citizens groups in the Buffalo Creek area had signed petitions in 1963 and again in 1968 asked that "the dam be drained periodically" but that nothing had ever been done.

Anger over the flood was all the greater because it was directly related to strip mining. In an interview, Gene Harner of West Virginia University said, "When you strip, you take away the outcroppings left by the deep mines. Then the water bursts through. Also, stripping destroys vegetation and leads to floods, particularly in the early spring before things grow."

A retired miner, Marry Barnett said, "As soon as my children come of age, I'm moving over to Virginia. They're just tearing up West Virginia." Another valley resident said, "What we need is another Ku Klux Klan to come in here. Only this time we'd blow up some of that heavy mining equipment."

Critical of the governor's investigating committee, some West Virginians set up another group. These were members of the Black Lung Association, an organization concerned with the disease common to miners. "The current members of that panel are either oriented to coal or apologists for the tragedy," said Pat McClintock of the group. "So we are creating our own commission to take testimony from eyewitnesses. We do not see this as a disaster in a vacuum but a series of events of coal dominating the lives of West Virginians."

Governor Moore ordered that 100 impoundments similar to Buffalo Creek be emptied. He also reacted to national press coverage of the disaster. "The only real sad part about it is that the state of West Virginia took *a terrible beating which far outshadowed the beating which the individuals that lost their lives took*, and I consider this an even greater tragedy than the accident itself."

At the time of the flood, John D. Rockefeller IV, West Virginia's secretary of state, was running against the incumbent, Arch A. Moore, Jr., for the job of governor. He toured the area of the disaster the day after it took place. He made no statement, however, until a speech several weeks later. Then he said, "I refuse to accept the theory that the flood was a 'natural occurrence' and that the impoundment was 'logical and constructive.' " (Governor Moore had said just these things right after the flood occurred.)

Rockefeller, who was also fighting for the total abolition of strip mining, went on to say, "We want the deep mining industry in particular to grow. But we also want it to be responsible. These are not acts of God. They are corporate and public negligence."

The Pittston Company had not yet issued any public statement but it did send a message to its shareholders in which it expressed its sympathy for the families who had suffered and then went on to tell how company officials had given assistance since the disaster. At this time it was disclosed that one of Pittston's directors was the brother of Secretary of Interior Rogers Morton, whose department had charge of the Bureau of Mines. No one tried to prove that this connection had any sinister meaning.

A month after the flood the death toll had risen to 118 persons. The Buffalo Mining Company had begun to accept claims against it but would not admit to any responsibility. No settlements had yet been offered. Some of the younger, angrier men, a number of them veterans of Vietnam, wanted to shut Buffalo down "until they pay us for what they did." Yet, a month later, the coal mining was nearly back to normal, with freight trains of the Chesapeake and Ohio railroad running day and night. That was about all that was normal in the valley. Piles of debris still burned at all hours. Many houses were still without water and many people had to use portable toilets set up along the road. At a community meeting, an older woman shouted, "We may not be alive in ten or fifteen years when the Surpeme Court gives us our money. We want Pittston to pay us now. They own the land up there. They put the water up there. They're responsible." Late in April the company paid its first claim; $4,000 to an employee for property damage. The matter of liability would eventually be settled in court.

On September 7, 1972, when the death toll had settled at 125, Governor Moore's commission gave its report. Commenting on the flood that also left 4,000 people homeless, the report said that the "Pittston Company cannot be excused from responsibility for faulty construction of the dam." It went on to say that the company had violated both state and Federal laws in building the dam. The report also accused the Bureau of Mines of "contributing to the causes of the disaster" by failing to enforce a Federal regulation against refuse piles that interfere with natural drainage. The West Virginia Public Service Commission was also charged with failing to enforce a law requiring permits for dam construction. In addition, Pittston was said not to have included the dam and two others like it on maps sent to the State Department of Mines only the day before the dam gave way. The report also addressed itself to the coal industry in general, telling it to "take a look at those practices which affect public safety and the environment" and to "explain publicly" why waste material cannot be better handled in worked-out mines.

Governor Moore, who had been said by some critics to be overly friendly to the coal companies, did not praise the report extravagantly but he did accept it and urged the coal industry to begin "a very deep searching of its own practices."

All this will not bring back the lives lost, repay the survivors for all that they suffered, nor do anything about the material cost of the flood. Yet the Buffalo Creek disaster may help in the campaign against strip mining. And, as will be seen, it has already had a profound effect on programs for much greater dam safety in the United States.

Dams That Go Bust

Collapsing dams are merely one cause of floods but perhaps the most dramatic because they come with such swiftness and usually as a complete surprise. In the United States the first major dam failure on record came in 1874, when a small storage reservoir on Mill Creek, a tributary of the Connecticut River, broke loose after a landslide along one bank. Water flowed at a rate of 60,000 cubic feet per second and the reservoir was drained in only 20 minutes. 143 people were killed and more than $1 million lost.

In 1889 came the dam disaster so ghastly that it has earned a permanent place in the national lore. To store water for operation of the Pennsylvania Canal, a barrier had been built in 1853 on the Little Cone-maugh River in Pennsylvania, 15 miles upstream from the small city of Johnstown. A body of water called the South Fork Reservoir was created and it had a capacity comparable to 11,000 acres of water a foot deep. The dam was about 70 feet high and 900 feet long. Competition from the Pennsylvania Railroad put the whole canal system out of business a few years later and the reservoir was neglected, forgotten. It began to leak at its base. In 1879 a club of wealthy Pittsburgh men bought the old reservoir to use as their private fishing and hunting property. A break in the dam was repaired and the old discharge pipes removed so that the reservoir would remain full. The dam was cut down by 2 feet so that it would be wide enough for a road; thus reducing the margin of safety. The spillway was partially clogged by trestles to support the road and by wire mesh to contain the fish.

Heavy rain began to fall at the end of Memorial Day in central Pennsylvania and it continued on May 31, 1889. The Conemaugh River began to rise and flood the houses and businesses in its valley. Just before noon on the 31st the South Fork Reservoir became full and water began to pour over the top. For three hours the flow scoured the dirt dam until it became too weak to hold against all the pressure from the huge amount of water it should have contained. It broke and within 45 minutes the reservoir was empty, its contents flowing down the narrow valley where 30,000 people lived.

A wall of water came plunging towards Johnstown, 15 miles away. It reached the city within an hour, destroying every building, railroad and bridge in its headlong path. The river, overflowing from all the rain, had already inundated the streets of Johnstown to a depth of 10 feet before the flood from the reservoir arrived. Escape would have been impossible for many people, even if warnings had been given.

All the debris the flood had picked up in its wild course, railway trestles, trees, houses, livestock, boilers and locomotives were caught by a stone bridge in Johnstown and this mass of jammed material created an island where innumerable people fled to save themselves. But this strange island caught fire and the refugees who survived the water burned to death. The estimate is that 2,100 people lost their lives due to flood and fire. There was national horror over the Johnstown flood and the careless owners of the reservoir were scourged. States, for the first time, passed laws about dam safety but these statutes did not prevent more dam failures in the future.

In West Virginia in 1914, January floods caused the Stony Creek Dam to give way. Winter storms in California in 1916 caused two dams to fail, the Sweetwater and the Lower Otay, with about thirty lives lost. Winter flooding in Southern California was so bad that year that San Diego had no communication with the rest of the state for almost a month.

Curiously, a very spectacular dam failure in California in 1928 seems unknown to almost everyone today. This may be due to the fact that there were absolutely no people surviving who actually saw it. The St. Francis Dam, 45 miles north of Los Angeles and just over the mountains from the San Fernando Valley, was made of concrete, 205 feet high. It had been built in 1926 to store water for the city. Two years later, on March 12, the Chief Engineer of the Los Angeles water works inspected the dam and expressed no concern over some cracks that had developed in the concrete nor over the fact that there was some leakage. About midnight that same day the dam burst and 38,000 acre feet of water went rushing down the San Francisquito Valley and out to the Pacific. More than 350 people were killed in its path and the damage came to about $15 million. The dam's caretakers were swept away in the flood and there were no eyewitnesses to describe what happened. The killer came and went in the stealth of the night, leaving only the appalling wreckage as mute testimony in the morning.

Investigation showed that the St. Francis Dam had been built on a fault, a crack, a weakness in the earth's crust subject to movement without warning. In addition, the dam had been set upon weak rock of a kind that gradually decays on contact with water. A dam with some elasticity might have held but not a rigid structure such as was actually put there. Since then more careful geological studies have been made before dams are built, yet California recently had a failure in a reservoir and a near disaster at another, due to geological causes.

Naturally, the United States is not alone in having dams collapse. In 1957, a dam of concrete with masonry buttresses was built across the Tera River in northwestern Spain to create hydroelectric power. A reservoir 2.5 miles long began to form, but the capacity of the dam was not fully tested until two years later, the winter of 1958–1959. That was a season of unusually heavy rains and the Tera rapidly filled the storage basin. Around midnight, on the night of January 9, 1959 water reached the crest of the dam, creating so much pressure against it that 17 of the 28 buttresses gave way. Almost immediately 300 feet of dam wall crumbled and a wall of water 200 feet high and 2.5 miles long came tumbling down a 1700 foot mountain slope into the unsuspecting village of Rivaldelago. The noise was incredible. It alerted several hundred villagers who managed to get up the hillsides for safety but 123 others did not make it. Nothing in the town of Rivaldelago remained standing.

The inevitable investigation showed that the dam simply was not sturdy enough for the work demanded of it. The engineers had miscalculated.

In France that same year, another proud new dam took even more lives. Built in 1954, the Malpasset Dam on the Reyran River was 200 feet high and of the most modern design. There had been no particularly heavy rains. One December day, the Malpasset Dam just collapsed and a vast river of water came rushing down on the little community of Frejus, a Riviera resort first founded by Julius Caesar. Engineers found that part of the dam's base had shifted on a soft foundation. Thus weakened, the whole structure finally came apart due to the weight of the water behind it.

RAMPAGE

The problem of safety was again brought to public attention on October 9, 1963 when water from the Valont Dam in northern Italy came down and wiped out almost 3,000 people. The dam itself did not break, however; the structure stood against forces much more powerful than had ever been planned for it. What happened was that huge portions of a mountain collapsed and millions of tons of rock and mud came down in a landslide that forced the reservoir to overflow. That the dam itself held was not much consolation to the victims; why had no geologist discovered the mountain's instability?

Two months later, the citizens of Los Angeles had their own shock. As described in a report published by the California Department of Water Resources,

> The day of December 14, 1963, began quietly. At 7:45 A.M. the caretaker of the Baldwin Hills Reservoir, Mr. Revere Wells, drove up the roadway from La Brea Avenue to the reservoir site. After making his routine readings of the inflow and outflow meters at the tunnel portal, as was his daily practice, he drove up the roadway to the caretaker's station on the east side of the reservoir. He crossed the footbridge to the gate tower and observed the water level in the reservoir. All the readings taken and observations made at the tower were regarded by Mr. Wells as normal at that time.
>
> At about 10:45 A.M., Wells prepared to make

a regular inspection of the spillway catch basins on the north face of the dam. Listening for a few minutes, he had the impression that the flow from the drainage system was increasing.

Mr. Wells immediately phoned to the dam officials, reporting that something was wrong. Even as he talked, water had started flowing out from the base of the earth dam, 150 feet high, and down the slopes of the Baldwin Hills which rise just east of the Los Angeles International Airport. The flow quickly became a flood as the water cut open an ever larger hole. Within three hours the break was seventy feet wide and houses were being torn apart and automobiles destroyed in a five-mile area below the hills.

The authorities had rushed to the scene at once, after Mr. Wells' call, and immediately realized that a disaster was occurring. The alarm was sounded. Police rushed into the threatened area, blowing whistles and sirens, pounding on doors, warning people to get out. From the time of the alarm, citizens had two hours of leeway before the worst happened.

Newspeople heard the word that a dam was going to break and rushed to Baldwin Hills with cameras. Many people in Los Angeles knew what was going on and were watching the event on TV. The dam split open early in the afternoon and a camera, operating from a helicopter, showed the awful event just as it happened. Had the Baldwin Hills reservoir given way in the middle of the night, a region housing thousands of people, who did not even know it even existed, would have been wiped out.

With California's meticulous inspection laws, how could this have happened? Baldwin Hills had been most carefully studied by geologists and seismologists before the first shovel was turned. The site was declared perfectly safe.

At the beginning, it had been. However, not too far away, there were a number of oil wells constantly pumping material out of the ground. Other wells were taking out groundwater. Over a stretch of time, these withdrawals had become quite significant. Support for the surface was being removed. Water, under the tremendous pressure of the reservoir, found small cracks and weaknesses and exploited them. Water will find its level. In this case, the action was hidden and did not reveal itself until much too late.

At almost exactly six o'clock on the morning of February 9, 1971, residents of Los Angeles County were wakened by the world's most forceful alarm clock, an earthquake. There was damage over a radius of 20 miles but the worst was near the base of the foothills at the northern end of the San Fernando Valley. Here, three hospitals were put out of commission and, in one of them, more than 40 patients killed when ceilings and walls collapsed. Freeways and overpasses were destroyed, many houses damaged beyond repair and more than a thousand people were injured badly enough to need some treatment.

But the worst disaster the earthquake might have caused never did happen. Within a mile or two of the ruined hospitals and much closer to the highways were two reservoirs of the Los Angeles water system known as the Upper and Lower Van Norman dams. They were dirt dams, built in the 1920s by a method, now obsolete, known as hydraulic land fill, to store water piped in from the Owens River Valley.

Some time before the earthquake, after one of the frequent inspections, it was decided that these Van Norman dams were really not as strong as the capacity at which they had been rated. They were kept considerably less than full and it was planned to strengthen them in the near future.

Moments after the shock at six o'clock the caretaker of the dams was out seeing what, if any damage, the embankments had sustained. He saw that the dam face of the lower dam, overlooking the San Fernando Valley below and where there were about 80,000 houses, had been smashed. The concrete facing was broken and the water was washing away the dirt underneath. The caretaker phoned his superintendent, who rushed to the scene, and by eight o'clock it was decided to try to empty both dams as rapidly as possible. The police were informed and, by nine o'clock, it was decided to use the powers they possessed and order all the residents below the endangered structures to evacuate their houses. The order was broad-

27

cast on the radio and police cars drove up and down the streets, that would have been flooded in case of disaster, shouting their instructions to the citizens with battery-operated bullhorns. The residents of the Valley left their homes and possessions behind and drove away from the area with hardly any complaints. They were not allowed to return to the area until the dams were at last declared safe four days later.

California has the most complete program for dam safety of any of the United States. How, in view of all their care, could the Van Norman Dams come so close to catastrophe? The answer is that the engineers' standards were wrong. These standards were based on past experience with earthquakes. One of these standards has to do with ground acceleration, lateral motion of the earth. The standard called for building structures that could stand up under a sideways shaking at .25 the force of gravity. The San Fernando shock confounded all past experience by shaking at more than 1.0 g, the force of gravity itself. Such acceleration had never been observed before.

C.J. Cortright, the engineer in charge of the Division of Safety of Dams for the State of California, wrote in a letter: "Obviously something must be and is being done about all this. More meaningful and representative methods of analyses recently developed are being applied to both existing and proposed new dams. Earthquake design criteria more representative of the anticipated ground motions are evolving. Selected dams in California are undergoing investigation. Dams identified as not possessing the requisite strength are being strengthened, replaced, removed or restricted in use. Many practical as well as political, social and economic considerations are involved in the process. Solutions will require time. None of the decisions will be easy."

As long as nature continues to produce surprises, dam building will never be absolutely safe but, in California at least, the officials are vigilant.

Big Rain on Rapid Creek

The morning weather forecast for Rapid City and the tourist region of the Black Hills in South Dakota on June 9, 1972 was "Variable cloudiness with a chance of scattered showers and thunderstorms through Saturday." The Rapid City *Journal* carried several enthusiastic stories about South Dakota Senator George McGovern's chances of winning the Democratic Presidential nomination. A large photograph, on page 3 of the newspaper, taken from the air, showed winding little Rapid Creek and the new East St. Patrick Street bridge half way through construction. The caption explained that work had been slowed somewhat by rainy weather during the past month.

Among the entertainments offered that Friday evening were several dances and dirt track racing at the Black Hills Speedway. "If rained out will be held Sunday at 8 P.M." The big event that night, however, was in the Fine Arts Auditorium at Stevens High School where a large audience gathered to hear a concert given by a visiting 70-piece high school band from West Germany. The program was only half completed when the director stood aside while an announcement was made that the rest of the concert was cancelled. Rapid City was filling up with water.

Two of the boys in the band were staying at the home of Dr. Lowell Dieters and they left the auditorium with the doctor and his family. Two hours later Dr. Dieters drowned to death while trying to save his family and guests. A daughter also died but the German boys survived.

Two light planes operated by the South Dakota School of Mines had been up in the air cloud-seeding that day. Not that additional rain was needed in the area. During June there had already been 1 inch more than normal and the ground was well soaked but the meteorologists had been conducting such experiments for six years and no one saw any reason not to continue even though atmospheric conditions were a bit unusual. (On Saturday the National Weather Service said that the conditions at the time of the storm were likely to occur only once in 100 years. Heavy rains are not unusual around Rapid City but usually they move off into the hills. This time they did not move and 14 inches of rain fell in six hours.)

The rainmaking planes had had little success and

were back down at the airport. The weather radar at Ellsworth Air Force Base, however, was spotting growing thunderheads advancing on the area. These observations were passed on to the Weather Bureau at the public airport and observers there foresaw heavy rain. An amateur observer who lived up in the hills some miles northwest of Rapid City phoned in to say he had measured four inches of rain in his district between 5 and 7 P.M.

With that, Derry Nuby of the Weather Service put in a call to the local radio stations to issue a high water warning. He also called the Civil Defense and the State Highway Department. This was at 7:15 P.M.

Mayor Dan Barnett of Rapid City had been having a very pleasant Friday afternoon. A round of golf and then, in the evening, a swim at the Y. As the Mayor was leaving around 8 o'clock an attendant remarked "I guess we can expect some big rains on the creek." This tipped the scales for the mayor, who had been becoming more and more concerned about the weather. From his official car he radioed the police department to have the director of public works, Leonard Swanson, meet him at city hall. Swanson was there and also the chief of police, Ron Messer. There, seeeing the latest weather reports, the men decided that the trouble spot was going to be Canyon Lake and the dam and they raced to that destination.

Rapid Creek, which runs through the city, had first been dammed in a natural bowl between the hills in 1890. This had created Canyon Lake, one of the local beauty spots until a flood ripped it out in 1907. Then, however, most of the land along the creek, below the dam, was still farm land and few people had built houses in the scenic area. Damage was slight.

Canyon Lake was not created again until the 1930s when the dam was rebuilt by the W.P.A. as a depression-born project. The lake soon became extremely popular for boating and fishing and for some of the best swimming anywhere in the Black Hills. The dam survived floods on Rapid Creek in 1935, 1952, and 1962. Meanwhile the flood plains below the dam were developed into a very desirable residential area, with many expensive houses. Five years before the flood of 1972, Canyon Lake had been drained and the spillways rebuilt but it was gradually filling with silt and there had been plans for doing something about this in the summer of 1972.

The officials from city hall reached the site of the dam at 9:30 the night of June 9. City crews had already been called out to try and keep the spillway free of debris. The mayor and the police chief first set up a command post at a cabin below the dam. There, on the mayor's radio, they heard calls of emergency; one about a heart attack at the dog track, the other begging for a boat behind Canyon Lake. Mayor Barnett tried to find the boat of a friend with whom he sometimes fished, but the friend was in the hospital and the boat gone. Returning to the cabin, he found that the police chief had moved to the Canyon Lake Club. While this was going on, Director of Public Works Swanson was working right along with his men trying to keep debris out of the spillway. They were having some success until a dock broke away upstream and headed for the spillway gates. The crew tried but could not break it up.

Swanson thought he had better see what was going on upstream. He found water going round the north end of the dam and water on the road up to the hubcaps of his car. Seeing this, he felt the dam could not hold and he radioed the crew, ordering them to get off the dam and on to high ground.

At about 10:15, the mayor received an emergency call from a man on a highway near Rapid Creek, northwest of the city. There was a "wall of water" coming down the creek and it would hit the lake in about 20 minutes. "That was enough for me," said Mayor Barnett. "I called the police department and directed that all areas along Rapid Creek be evacuated immediately. I then called KOTA radio and asked them to relay this message to the people."

From then on that night the mayor's activities were heroic but a little hard to follow. For a time he rushed from house to house along the creek, warning residents to leave. The National Guard was fortunately in summer encampment. They had been called into emergency duty and the mayor had a number of them also hurrying from house to house, issuing warnings. But, whether from fear or disbelief, many people would not heed the alarm. "They wouldn't listen," the mayor said. All the rest of the night Mayor

Barnett hurried around Rapid City, sometimes making personal rescue efforts, at others coordinating information or calling for more assistance.

The 500-foot wide dam broke sometime between 10:45 and 11:00 P.M. Then the horror began for thousands of people.

Freed from the harness of the dam, vast quantities of water rushed down the valley, high above the normal banks of Rapid Creek. "It was a terrible roar," Mrs. Ray Stevens said," just like the Colorado River. People were screaming and shouting, and junk and houses were floating down the street," Mr. and Mrs. Stevens and their four children tried to escape by car but it filled with water before they had driven a block. The Stevens finally reached the top of a nearby house where they stayed all night with 10 other poeple until they were rescued early in the morning. Teresa Stevens, a teen-ager, remembered, "There was one man on the roof with us and he saw his wife across the street. She was screaming for him. I heard her shout, 'Oh, my God'. Then she was swept away. He yelled for her all night long. It was terrible. I thought it was the end of the world."

Dorrance Dussek went to help a crippled neighbor. Wading through waist-deep water, he found two boys trapped in a house nearer his own. He helped them out and then was himself caught in a current and swept along by the water for two blocks until he managed to grab a corner of a house. There he held himself until the owner pulled him inside.

Mrs. Gertrude Lux, aged 71, small and apparently not very strong, shared a house near the creek with her granddaughter, who was both physically and mentally retarded. Alone with the 16-year-old girl, when the water reached her chest in the bedroom where both were trapped, Mrs. Lux thought, "If it goes any higher, this will be it." She knew the girl was not strong enough to stand in the water but a foam rubber mattress had floated off the bed and she got Vicki to lie on it. The mattress simply rose as the water did. Mrs. Lux was afraid the girl would tip the mattress and roll off, so she stood there balancing it. Standing was difficult because the water was rushing by and the footing "was all so slimy."

Next door neighbors, the Charles Berrys, and a visiting friend, Chet Andrews, had earlier urged Mrs. Lux to take Vicki up on the roof but Mrs. Lux knew Vicki could never climb to the roof and she thought she would be no safer up there in the rain and the dark. Andrews broke the windows in Mrs. Lux' house, so that the water could not rise higher outside than inside. "It saved our lives." Andrews then went up on Mrs. Lux' roof in hopes of saving himself.

During that long night, Mrs. Lux stood in the cold water, fearing it might rise higher, fearing fire, but protecting her helpless granddaughter. The girl herself even slept for a while. Toward dawn the water began to go down. During the night she had heard the sound of voices and boats but then they would disappear. The current was too strong for those trying to rescue them. Mrs. Lux and the girl waited with Chet Andrews until 11 in the morning, in the house full of mud and water, until help finally came. The girl and the brave old woman were taken to her daughter's home.

"I looked out the window," Mrs. Joseph Ellis remembered, "and I saw him in the flashes of lightning. He was caught in a tree over there. He was a brave youngster, only about 10. I kept hollering at him to hold on and climb higher. He answered 'I'm above the water and I'm holding on real tight.' "

Her son would yell to the boy "Keep the faith, buddy," and with each flash of lightning the people in the house could see more people trapped in the mad jumble of floating trees, houses and cars.

After midnight, the water went down somewhat and Mr. Ellis could see the top of a car. "I figured it was down to about five feet" so the Ellises tied a rope on to the front porch and, hanging on to it, went to rescue the boy with the help of two National Guardsmen. Later that night they also saved a man they saw clinging to a telephone wire and a terrified dog floating by.

For themselves, the Ellises know they were lucky. The house next door floated a quarter mile down the street. Sometime during the night another house ended its journey in the Ellis back yard. Three cars also became wedged against a row of trees in their yard and they believe that the other house and the cars kept the force of the water away, allowing their house to stand. The boy whom the Ellises saved found his father at a rescue center the next day. Then father and son began to search for the rest of their family.

When the flood waters drove them out of the first floor and then out of the second, Mrs. and Mrs. Harold Bruns went up to the attic for safety. There they discovered inner tubes their daughter had once used for floating down Rapid Creek. They had just gotten the tubes under their arms when the roof exploded and they were thrown clear of the house. They lost each other immediately. Mrs. Bruns managed to find footing on the top of a car that had been stopped by a tree on the golf course. Mr. Bruns succeeded in climbing a tree after his inner tube broke. Both were rescued but not until late in the morning did each discover that the other had survived.

National Guardsmen, while driving a truck carrying people to high ground, reported seeing a house floating down stream with a number of people on the roof. They could only watch, helpless, as the house struck a bridge and threw all the people into the water. Moments later they saw a young man and woman on the roof of a car floating by. Then the car struck the same bridge and the couple was also spilled into the torrent.

Sam Chase was a sturdy young man of 21, but he had to hold on to his father's arm as his father led him to the health department for typhoid and tetanus shots. "I couldn't help it," Sam did not quite cry. "My car was swept away. I was under the water. I got bruises and cuts." He asked softly for a tranquilizer. His father said then, "His wife was one of the ones . . ." and then said no more.

When the family of Davis Heraty heard Mayor Barnett on the radio Friday night warning about the flood, "We thought he was kidding. We just sat there and pretty soon this big bunch of water came down the creek. We ran next door and suddenly the water was up to my neck. The top of a house came floating by and we grabbed on to that. A little way downstream we got off and climbed on the roof of a neighbor's house. We stayed there until the flood began to fall on Saturday."

After Kent Larson saw the flood waters break down the doors of his garage and carry his car out the back wall, he decided to open the glass doors in the living room of the house he had recently bought. The water flowed through the garage, then across the living room and out past the glass doors. The only damage to the house was a four-inch carpet of mud on the floor after the waters receded. The car was later found fifteen blocks away, completely smashed up.

Six boys who were celebrating their graduation from high school in St. Paul, Minnesota had been camping out in the hills but decided on Friday night because of the rain to take a room in a motel in Keystone. The motel was right on the edge of Battle Creek, a tributary of Rapid Creek. At a quarter to 10 the electricity went off and the tired boys decided to go to bed. A little before 11 P.M., Tom Doherty heard water coming into the cabin and he woke up his friends. They tried to open the cabin door but the pressure of the water kept it shut. Then Mike Kovacovich kicked out a window, moments before a car smashed into it. Then the youths grabbed mattresses and floated on the water, while the cabin itself started floating down the stream. "It went at least a mile" Mike Kovacovich said, "and then one wall of the cabin broke away from the rest of it. I'd given myself up for dead. I thought this was it."

By this time all six were floating free of the cabin. The mattresses became waterlogged and sank. William Gange and Kovacovich succeeded in getting a hold on a building as they floated by. Bruce Glover grabbed on to a gas station farther down the rushing stream. Kovacovich and Gange got to the roof of the building and sat there in the cold and dark; Kovacovich wearing only a pair of shorts. There were a number of people who had found safety in a solid building next door and they finally saw the plight of the two boys. A plank was maneuvered over to the roof top and the two crawled to safety. Next day in the town of Custer they were treated for bruises and cuts—and then began to look for the bodies of their three dead friends.

The day after the flood, Tom Murphy, 22, went to look at the ruins of Canyon Lake Dam. "There used to be trees on it, and everything. It's sort of unbelievable, if you'd seen it before, to look at this." He motioned to the gap in the barrier, 500 feet wide. "There used to be houses done there." Had Murphy lost any friends? "Yeah, some. They lived right in the path of that." He pointed to a horrible mass of uprooted trees, pieces of the dam structure, twisted cars. "My girl-friend lived there. She was living there with her in-valid father. She'd have had a hell of a time getting him out."

Murphy had been out of town the night before and could only manage to get back to Rapid City on Saturday afternoon. His own house was on a hill and suffered no damage.

Tales of that awful night kept coming in for weeks. In Washington, D. C., on June 25, Congress-man Jack F. Kemp of Buffalo, New York received a letter from Norman Baker, a history professor at the State University in Buffalo.

Dear Mr. Kemp. I have lived in Buffalo for three years and although, as an Englishman, I have not voted, I hope you will not mind my regarding you as my representative and asking your help. We recently had the misfortune to be in Rapid City, S. D. on vacation when the flood hit. We spent some time chest deep in water holding up our two children with only a small tree for protection from the force of the flood and the danger of de-bris. Our strength was virtually spent when six men from the nearby Air Force base came through 200 yards of water at great personal risk to rescue us. In particular, the leadership of Cap-tains Soll and Knutson were instrumental in bringing us to safety. All six then went back into the water and saved others.

I have written to Colonel Reed at Ellsworth AFB, the men's commander, sending a check for an Air Force charity, and asking some official commendation for these men; if possible, the Air Force Medal.

Professor Baker explained he was writing Congressman Kemp to help see that "these men receive the commendation they deserve." Congressman Kemp proudly read the letter into the *Congressional Record*.

Aftermath in
South Dakota

Dawn finally came after the midnight rampage, grim and hideous. Civilian volunteers, men from the Air Force base, more than a thousand men from the National Guard, most of whom had been working half the night, continued along the path of the flood, looking for survivors, digging into the wreckage for bodies. No one knew how many of Rapid City's 43,000 people had died, but at least 80 blocks of buildings along the creek were wiped out. It was a strip 4 blocks wide and 8 miles long.

Organizations such as the Red Cross, Salvation Army, and Civil Defense set about their tasks with the efficiency born of planning. They dispensed first aid, food, and clothing, and set up centers of information to help survivors find their relatives and friends. By midmorning a major problem developed when the Bennett-Clarkson Hospital had to evacuate all patients. The utilities were completely out (as they were all over Rapid City) and even though a National Guard truck brought a tank of water to the back door, this was not enough for sanitation in the big hospital. The laboratories in the basement were completely destroyed. Doctors and nurses had struggled to keep Bennett-Clarkson going but finally had to admit defeat. Patients were moved to the remaining three hospitals in the city. The hospital had had the only obstetrical service in the community and one woman was in the process of delivering a baby while the move was underway.

Radio Station KKLS was flooded out at 11:34 Friday night when a wall of water broke down the door. Just before midnight the KOTA television transmitter was struck by lightning and just after midnight KOTA radio went off the air due to loss of power. KOTA radio resumed broadcasting at 4:14 Saturday morning after it obtained portable generating equipment. Radio station KIMM, usually restricted to broadcasting between sunrise and sunset, was put on the air at 4:30 in the morning and thus citizens with battery radios, at least, had some contact with the world. Electric power was gone, of course, the telephones were out, and the water shut off until it could be tested for safety. Typhoid shots were urged for everyone.

Late that Saturday a panic nearly broke out in

Rapid City when a rumor spread that the enormous Paciola Dam, 25 miles from Rapid City, up in the Black Hills, had broken. Some witnesses said it all began when a man dressed in a National Guard uniform walked into the Meadowbrook Grade School, crowded with refugees, and shouted through a megaphone that Paciola was gone and that everybody should flee to high ground. With this news, several carloads of people began driving around the Meadowbrook area, shouting, "The Paciola Dam is broken. Get to high ground." Many people rushed to their cars and began driving off. Within moments, a huge traffic jam developed at an intersection in Meadowbrook. Many people did not believe the rumor and tried to talk their neighbors out of panicking. Finally, announcements over the radio that the dam was perfectly safe restored public order.

But in Sturgis, South Dakota, 27 miles from Rapid City, the danger from another dam was quite real. This dam had been built by the Army many years before to supply water to Fort Meade but the Army camp had been converted into a Veteran's Administration hospital and the reservoir was no longer in use. Still, it was left standing. The Friday night rain that struck Rapid City also fell around Sturgis and the erosion it caused through an abnormally swift overflow on the front face of the concrete and dirt structure washed away huge slabs of earth. A torrent rushed down from an elevation of 4,000 feet, washed out bridges and roads, and then crashed through

Sturgis. There were no injuries in the town of 6,000 people but hundreds of basements were flooded and 20 houses destroyed.

Patients from the local hospital were evacuated to the V.A. Hospital which stood on higher ground and older persons who lived in the valley were urged to find refuge above the water's possible reach. Other residents, hearing this news, began to pack and leave town. An official broadcast was made to reassure people that the moves were only precautionary but that "additional rain could cause the dam to breach." People were urged, however, to pack suitcases and be ready to move on a moment's notice. Departure routes were announced. The signal to leave would be two long blasts on the town's siren.

Meanwhile, pumping teams were sent to lower the 55 foot level of the dam, which was holding back 20 million gallons of water. The Air Force flew in additional pumps from all over the United States and these were dragged up the steep mountain roads to the site of the dam. The work went very slowly on the badly damaged structure. An Army Engineer said, "Nobody knows why it's still standing." Nevertheless, the level of the Fort Meade dam was lowered sufficiently before the next serious rain and the town of Sturgis was saved.

In Rapid City, the roster of the dead kept changing but the number was over 200 and it was certainly the worst disaster in the history of South Dakota. As of June, 1972, it was the greatest flood in the United

States since the Ohio and Mississippi rivers took a toll of 250 lives in 1937.

As the city struggled to pull itself together, a number of predictable events took place. Immediately after Washington received the news, Rapid City was declared a disaster area and the President sent his aide, Robert Finch, to look at the scene. Later, Mrs. Nixon came to a memorial service for the victims. The governor of South Dakota was also at the service. Senator McGovern interrupted his campaign for the Presidential nomination to fly to his home state and surveyed the scene of the flood two days after it happened. Later, after the Democratic convention, McGovern and his wife took a vacation at a Black Hills resort and urged Americans to visit the region that summer as a means of helping. Less than one percent of the tourist facilities had been damaged. Partially to advertise the beauties of the Black Hills, McGovern sat for a national TV interview outdoors at the Sylvan Lake Resort in Custer State Park.

Curiosity seekers, even from far away, arrived to have a look at the devastation, clogging the already crowded streets so much that police were ordering, "Move on or be arrested." The roads had been cleared for the use of heavy equipment, needed to get rid of the debris, and hot sunny days turned the mud into dust so that every passing truck raised a thick cloud as it passed. Electric and telephone service was mostly restored by the fourth day but water was still in short supply. Some people tried to wash their dusty cars and, those who still had them, to water their lawns but Mayor Barnett quickly issued an order against such use of water until it was generally available again.

When the first check was paid out for relief by the Small Business Administration, in the record time of seven days, the event was given maximum publicity. Seventeen-year-old Michael Coyle and his sixteen-year-old, pregnant wife, Debbie, were interviewed as they received a $2,500 loan and the story of their particular night of horror was given half a page in the local newspaper.

Almost as soon as the news of Rapid City flashed across the nation, hordes of reporters and photographers began to appear, even two men from London newspapers. Energetic Mayor Barnett became something of an instant hero. He testified before Congressional committees on flood relief legislation, flew to several cities to accept checks for flood relief, and went on national television shows at least twice. He was somewhat defensive about the city-owned dam. "It was a very good, sound stable dam for a normal river." It had stoutly held the first rush of water. "The second burst of water was more than we could stand." The mayor blamed much of the damage on improper zoning, building on the flood plain. He announced that the city would be wiser in the future.

Many people who had lived in the lovely, tree-filled area near the creek wondered if, in any case, they wanted to rebuild there. Don Carrier, who had bought a house there three months before the flood, had been lucky in that he took his wife and child out for a soda at 10:30, before the flood struck. The family spent the night in their car after learning of the event. His house, of course, was gone. "People always say this is the high-class area. Well, I tell you, there are some other parts of town that are looking pretty good about now. Nobody's going to sell me a house near a river again!"

In the aftermath of all the heartbreak, two questions remain unanswered. Will the people of Rapid City keep their good intentions? Will they turn the banks of Rapid Creek into public parks and resist the temptation to build again in this attractive place?

More generally, the dam that made Canyon Lake had not been neglected. It had been repaired only five years before it broke and was scheduled for another renovation. If, as the mayor said, "It was a very good, sound, stable dam," what of others like it? Unusual rains may happen anywhere and put "safe" dams to unusual tests. In just more than a year, the United States had two major dam failures and two near dam disasters. What comes next?

Rivers of Sorrow

The most famous flood in history, according to the reckoning of the late Bishop Ussher in Ireland, happened in the year 2448 B.C. The Bishop believed that the Lord made the earth in 4004 B.C. and from the genealogies in the Bible he arrived at the date of The Flood. Archeologists and geologists for many years have been trying to find evidence in the Mediterranean for this flood, the story of which they believe must have had a basis in historical fact. So far no proof has been found but certainly the region is no stranger to floods.

Recent deep sea drilling by the research ship *Glomar Challenger* in the Mediterranean has shown that 5 million years ago for a period of some length, what is now the ocean floor was then a dry desert. The drill-

ing does not show how long this condition existed but at some time the sea was certainly refilled. Perhaps a dramatic upheaval at the Straits of Gibraltar opened the flood gate to an inflow from the Atlantic, a flow which continues today. If the flow came suddenly, it would no doubt produce a flood, but this may have happened a million or more years ago. Primitive ancestors of man would have observed the event. Could a race memory persist for such a length of time?

Records of floods on the Nile, which empties into the eastern Mediterranean, go back thousands of years. The Egyptians were not dismayed by the annual overflow of the river but took great advantage of it. The high priests would make fine calculations of when the flood would occur, but surely the peasants, as well, knew the approximate date from experience. They would pack up their belongings and animals and move away, to live in tents, while the waters would cover their farms and then retreat, leaving a new layer of dirt to enrich the topsoil and groundwater underneath the surface which could be recovered by shallow wells and used for irrigation during the dry months. Living in rhythm with the floods, several crops of wheat and other grains could be harvested every year.

This system was not good enough for modern man, however. The British who ruled Egypt at the beginning of this century found the floods untidy and in 1902 they built a dam near the first cataract of the Nile at Aswan. This did provide a measure of flood control but was still not considered sufficient for irri-

gation purposes. During the 1960s the Egyptians, with Russian help, built a much greater dam at Aswan, in the process flooding many antique ruins of great value. This dam, by depriving the Mediterranean of an important source of fresh water full of nutrients, has virtually eliminated sardine and other fisheries, increased the salinity of the sea beyond the tolerance of many species, and decreased the ability of the Mediterranean to digest all the pollution dumped into it by the many nations along its shores. As always, tampering with nature has side effects which often are not expected.

The worst floods in history have been caused by China's Hwang Ho, which we call the Yellow River, also known as "China's Sorrow." Since the oldest flood in China's records, in 2297 B.C., the Yellow River has overflown its banks more than 1,500 times. Rising in the mountains of Tibet, 2,900 miles from the sea, the river flows through lands that have been almost completely deforested for more than a thousand years. There is nothing to stop erosion or hold the rainwater, so the river carries away more than a billion tons of silt every year. Much of this is deposited on the flood plains and this is where the Chinese must do their farming, whether they endure floods or not. To keep the floods in check, dikes have always been built but they must continually be built higher since the dirt-choked river deposits keep raising the level of the river bed.

Among the river's notorious activities is its propensity for changing course. On one occasion the mouth of the river changed by a distance of 250 miles. It is as though the Mississippi suddenly deserted New Orleans and began to flow past Houston. Until this century the greatest Yellow River flood came in 1887 when more than 900,000 people drowned in its dirty waters. In this century its worst floods have been created by man. To slow the advance of the invading Japanese armies in June, 1938, Chiang Kai-shek ordered his troops to break the dikes of the river near the capital of Honan province. This worked for a short time but at the cost of 890,000 Chinese dead and 13 million made homeless. The present Chinese government has given major priority to strengthening the dikes of the Yellow River and there have been no reports of great floods in recent years.

Bordering China to the south is northern Vietnam, whose major waterway is the Red River which originates in the mountains of Central Asia. The delta of the Red River is a fertile plain where millions of Vietnamese farmers live but it is also a flood plain. When the monsoon winds, loaded with moisture, hit the mountains in late summer the rains flow into the Red River and its tributaries and the water, again full of silt because the region is virtually treeless, races downhill towards the sea. Archeological evidence shows that dikes were built along the river's banks at least 2,000 years ago although the first written mention of them is found in Chinese chronicles of the eleventh century. Like the Yellow River, the Red keeps depositing mud and thus raising its level until now the river bed is five or six feet higher than the fields around it. Thus, to battle the river, higher and higher dikes must be built and the warfare can never cease. (The same thing happened in ancient times along the Tigris and Euphrates Rivers until at last the dikes could be built no higher and the rivers finally won.)

In 1971 the Red River of Vietnam broke through a 30-mile stretch of the dikes and much of the fall rice crop was destroyed. In 1972, U. S. planes were bombing North Vietnam and the country's government claimed that the bombs were deliberately aimed to destroy the dike system. The U.S. Government denied this charge; but admitted that if there were a road on a dike or a gun installation, the dike might be damaged while the bombs were dropped on a legitimate target. In any case, the use of floods as a weapon of war is not new. During World War II, the Dutch opened their dikes in an attempt to repel the invading Germans and, later in the war, Allied planes bombed German reservoirs.

Along the coasts of India and Pakistan floods may come from two directions. In what was East Pakistan and is now Bangladesh, millions of people live on the deltas of the Brahmaputra River, which flows out of the central Asian mountains. The Brahmaputra frequently floods the lowlands and did so in 1971. Shortly thereafter, these lands which are barely above sea level were again hit but this time by the sea which swept over them during a cyclone. Relief from this disaster was so badly bungled by the dominant government of West Pakistan that the severely damaged province rose in a successful fight for independence. East Pakistan had had six such storms from the Indian Ocean in the previous 12 years with the number dead estimated at 75,000.

Floods on islands in the western Pacific are most frequently due to typhoons. Japan and the Philippines get these with unpleasant regularity.

All the famous rivers in Europe have overflown at one time or another; the oldest on record being the Tiber at Rome in 413 B.C. The cause was probably nothing more exotic than excessive rainfall but in recent years, in Italy, floods have been enhanced by overgrazing, overcultivation, and virtual elimination of trees. Yet the Italians have invested heavily in flood control. Under Mussolini, the Po River was thought to be under control and the works along its banks were highly praised. Yet in 1951, a combination of heavy rains and high tides, which prevented the river from emptying, caused the levees and dikes to break, driving thousands of people from their farms on the supposedly safe flood plains and killing 100 people and thousands of cattle. In 1966, the presumably tame Arno River ran wild through the beautiful city of Florence and destroyed many of the innumerable art treasures stored there. Heavy rains that spring also caused floods in Naples, Rome, and Venice. Venice, so close to the sea, is always in danger and in February, 1972, the famous St. Mark's Square was under three feet of water, driven there from the sea by high tides and strong winds.

Flood records for the Danube go back to the year 1000 and the river reached its greatest height ever in 1501. Holland's most lethal flood came from the sea in 1228 when something like 100,000 people were destroyed. The Netherlands were again devastated by a ferocious winter storm in 1953 which drove the North Sea over the dikes, flooded low-lying land, killed 1,700, and caused property loss amounting to millions of guilders. Records for the Seine at Paris date from 1649 and its greatest flood came in 1658.

No continent except Antarctica, has been without its floods. Gross destruction of the forests in Australia, to create grazing and farmland, have made its greatest river, the Murray, into a stream which alternates yearly between floods and extremely low levels. In the list of "Principal World Floods" in the almanacs, Brazil achieves status with a flood in Rio de Janeiro in 1966 which left 50,000 homeless due to high water and landslides resulting from heavy rain. Four hundred and five people dead. The following year Rio de Janeiro and Sao Paulo again made the list with floods that killed 600 people.

In North America, Mexico is increasingly subject to floods. These have always been expected along the eastern coast, exposed as it is to hurricanes, but the interior with its light rainfall nevertheless has periods of excessive water flow due to the heavy cutting of the forests. In the United States, of course, with all its varieties of topography and weather, there are floods somewhere almost every month and in every year. More than 10,000 floods can be listed for the United States from the time the first Europeans settled here. Many more no doubt went unobserved in areas that were not yet occupied.

For the earliest days, naturally, the records are quite meager. It is known that New Englanders had their first experience with the effects of a hurricane in 1635. The explorers Marquette and Joliet found a flood at the mouth of the Missouri River in 1673. At least twice in that century the Delaware River overflowed and so did the Connecticut. In 1735 New Orleans was inundated, by no means the only time. The Ohio River flooded in 1762 and again in the following year, measuring 41 feet above normal at Cincinnati. The first observed flood on the Poiciancia River, now the Los Angeles, was noted by a Spanish padre in 1770. The same year the Kennebec and Androscoggin rivers in Maine, the Connecticut in Massachusetts and the Chattahoochee in Georgia all overflowed their normal banks, probably due to a hurricane. The next year every river east of the mountains in Virginia flooded and the James River at Richmond took 150 lives. The Ohio flooded again in 1772 and 1773, rising 75 feet at Cincinnati.

The Red River which makes the boundary between what are now Minnesota and North Dakota, the only important river in the United States that flows north, first revealed its continuing problem during a flood seen in 1776. In the spring, as the warm weather advances toward Canada, the snow and ice melt first in the south, at the river's source. This spring-thaw water flows north and encounters ice where the river is still frozen over. Denied its normal bed, the Red River overflows on the flat lands of the two states. Continuing north with spring, on its way to Lake Winnipeg in Canada, the river repeats the same performance and floods the province of Manitoba. This behavior continues to the present day and it seems there is little that the wisest engineers can do about it.

During the remainder of the eighteenth century there were floods of particular note in Georgia, on the Connecticut, Mississippi, Ohio, Susquehanna, Merrimack, Kansas, Delaware, Lehigh, Tennessee, Cumberland and Yazoo rivers.

The Sacramento River in California was first observed flooding in 1805, a performance it has repeated a number of times since. A hurricane hit Charleston in 1813, and great hurricane waves caused the ocean to sweep over parts of Long Island and Connecticut. The Texas coast had a hurricane in 1829 in the unusual month of September. The pattern is supposed to be that these storms move to the Gulf of Mexico if they happen early in the year but aim up the Atlantic coast if they are late in the season. As the country gradually filled up in the years before the Civil War it began to be clear that no twelve-month period could pass without a major flood somewhere. During that war, in 1862, the new state of California had its most epic flood. Beginning in December, 1861, and lasting through the next January an average of 50 inches of rain fell everywhere from the Klamath River on the north to the San Diego River on the south. (The average for these two months in Los Angeles is 5 inches.) The valley of the Sacramento River was a great lake, and Los Angeles all the way from what is now downtown out to the sea was entirely under water. There were floods on twelve California rivers that January. The month before Lee surrendered to Grant, western New York and Pennsylvania had a mammoth flood and the Allegheny River broke all former records.

The following year the Mississippi River swamped a good part of North Dakota and the Miami River in Ohio flooded in the unlikely month of September.

One of the ten largest floods on record came in the spring of 1881 in the region of the upper Mississippi and upper Missouri Rivers. The Dakotas, Minnesota, Wisconsin, Iowa, and Nebraska all suffered damage from the waters such as had never been seen in that region before. Then in September of that same year, Minnesota and Wisconsin suffered through another flood.

The Ohio, which had not seen any flood records for 4 years, made up for it in 1884 when it reached heights in many places that still have not been equalled. This particular Ohio flood was worse than most because a sudden spring thaw over all the eastern

United States caused the winter snow and ice to melt simultaneously everywhere and all its numerous tributaries emptied into the Ohio at the same time.

The Gulf Coast of Texas was inundated by hurricane waves in June, 1886 and then again in August of the same year. The epic Johnstown flood came in 1889, as has been mentioned. Hurricane floods in August, 1893 took more than 2,000 lives along the southern Atlantic coast. In those days, of course, a warning system was hardly in existence. Next year the Columbia River staged one of the major floods in U.S. history, in terms of volume of water and amount of land involved. Floods on the Columbia are hardly rare but this one set records because, after a winter of unusually heavy snowfall, the snow barely melted at all until warm rains began to fall in May. Then the snow cover turned to water all over Oregon, Washington, Idaho, and Montana simultaneously.

During the five years remaining in the nineteenth century, major floods were noted in New York, Maine, New Hampshire, Wisconsin, Kansas, New Mexico, Maryland, New Jersey, Idaho, Montana, Missouri, Minnesota, North Dakota, Nevada, Utah, Ohio, Iowa, Oregon, North Carolina, Texas, and Wyoming. Yet, from reading today's newspapers, one would get the impression that each flood is a great surprise. There is probably not one stream or river in the United States that has not flooded at some time or another.

The twentieth century arrived with a million dollars of flood damage in eastern New York and New England during February of the first year. The biggest news, however was the Galveston hurricane of 1900. Winds and huge waves struck the Texas city on September 8 and killed more than 6,000 human beings. Damage was counted as $30 million. Soon after this disaster, the survivors built a sea wall to protect Galveston. This has served well enough so that another such storm in 1915 merely flooded the business district without taking any lives.

Routine, big floods occurred on the Willamette River in Oregon during January, 1903, on the Minnesota River in March and on the French Broad in Tennessee in April. May and June of 1903 were the months some people in Kansas still remember. Between May 16 and 31, 15 inches of rain fell in the area of the Kansas and lower Missouri Rivers. There were tornadoes in Iowa, Nebraska, and Kansas during this time that killed 50 people. By June 1 the Missouri at Kansas City had risen 35 feet. This was 14 feet above the danger line and only 2 feet below a record set in 1884.

In the countryside hundreds of farmers were made homeless and thousands of acres of corn destroyed. In Kansas City the platform of the Union Depot was under 6 feet of water. Wreckage destroyed every bridge across the Kansas and Missouri Rivers, 9 of them, and worst of all, the bridge which carried the 36-inch pipeline from the waterworks and which supplied the entire city. Damage was reckoned at $22 million and over 100 people were killed. Almost all the animals in the huge Kansas City stockyard were drowned.

Six days later, in the Carolinas, heavy rains caused floods which took more than 50 lives. On the 14th of June a flash flood in Oregon, caused by rain that only fell for half an hour, wiped out one-third of the town of Heppner, population, 1,400, and drowned something more than 200 people. A wave of water 25 feet high had washed down Willow Creek. In August of that year, torrential rains caused record floods in southern New York and northern New Jersey. Then, during three days in October, New Jersey took a further beating with 15 inches of rain. The Passaic River has never been so high, before or since, and 196 acres of the city of Paterson were flooded. Every bridge on six New Jersey rivers went down. Rivers in New York and Pennsylvania also flooded.

Northern California had had very severe spring floods in 1862 but because the channels were clear and the flood plains hardly occupied, the swollen rivers reached the ocean without doing much damage. Heavy snow and rain in 1904 over the same area found much of the area under cultivation and streams below the Feather River blocked by debris from mining. This time the levees broke at Sacramento and the capital of California was swamped. After this, the state government began to take steps to protect the rich farm lands along the San Joaquin and Sacramento Rivers.

RAMPAGE

In March, 1904, the Susquehanna caused about $5 million in damage at Wilkes-Barre. Michigan also had a record flood that month and the Wabash in Indiana drowned out a number of small towns. In September, the town of Trinidad in Colorado was nearly wiped out because of excessive rain.

The lower Colorado made history in 1905 and permanently changed part of the California landscape. In 1900 an irrigation canal had been built to divert some Colorado River water to the fertile but dry Imperial Valley. This canal soon began to be clogged by the heavily silted stream. Work was undertaken to clear the canal bed. The canal, notorious for its flash floods, had never flooded in any February on record and so this was the time chosen for the project. The Colorado did not behave according to the rules, however, and began flooding the man-made canal. This was all the easier because the river had earlier changed its course and flowed in this direction on five occasions in the previous century. There were no gates or barriers to prevent the Colorado from flowing down the canal in February, 1905 and the river did so, emptying its entire content into the Salton Sink, a geological formation 200 feet below sea level.

In time, the Colorado would have swamped the prosperous Imperial Valley, which is really just a part of the Salton formation. The situation was desperate but many people felt it was hopeless to try and control the powerful river. Yet E.H. Harriman, who controlled the Southern Pacific Railroad, a line which got much of its prosperity from the region, undertook to do something about it. Works constructed to put the river back into its original course were often broken down by successive floods and for two years the Colorado continued to flow into California until the sink became known as the Salton Sea, a body of water covering 500 square miles. When the river was finally thought to be under control in 1906, the works were again destroyed by a late winter flood. This time, however, the engineers had enough experience and they managed to close the gap in February, 1907. The project had cost the Southern Pacific Railroad $2 million but it saved the Imperial Valley, the source of out-of-season fruits and vegetables for all the United States. A bill was soon introduced in Congress to reimburse the railroad for performing a public service. The bill was turned down, however, because it was called "a gratuity to private enterprise." The Salton Sea remained and is today a popular recreation area.

The Ohio and Sacramento Rivers both overflowed in March, 1908 but the big floods began in May. Minnesota's three important rivers, the Red, the Mississippi, and the Minnesota all went wild and caused more than $20 million damage in the state. The Missouri also became swollen and caused trouble at Kansas City. Then all this water flowed past St. Louis, creating havoc, and the lower Mississippi remained over the flood stage for more than 100 days. Also, in May, four rivers in Texas began to flood, due to heavy rains. Five thousand people had to flee their homes in

Dallas and 600 people were driven out in Fort Worth.

The drought year of 1910 was notable in that only four floods occurred worth mentioning. In 1911 the Sacramento had its usual spring floods. The worst flood in memory came in October along the border between Colorado and New Mexico. Wisconsin had $3 million worth of damage at the little town of Black River Falls. Next year the Mississippi showed its power once again with destruction rated at $70 million.

1913 ranks high in the annals of floods due to the Ohio River whose misbehavior surpassed everything in its known history. Although the floods on the Ohio came in March, they were not due to spring thaw. The snows had all been melted by previous rains but the ground was already thoroughly soaked when two storms, one right after the other, struck the entire Ohio Valley. Rainfall was at least 10 inches everywhere but greatest in Indiana and Ohio.

The Miami River at Dayton rose to 29 feet above normal on March 26 at one in the morning, 8 feet higher than had ever been marked before. The Scioto and the Muskingum reached their crests at the same time and all three rivers poured their flood crests into the Ohio on the same day. Crests followed from the Monongahela, the Wabash, the Tennessee, and Cumberland Rivers. All this water, of course, had to be absorbed by the Mississippi, which was in flood during much of April.

The worst damage came at Dayton and the

smaller communities near it in Ohio. Known dead came to 467. Damage amounted to more than $100 million. The Miami River had seriously overflowed four times since the first pioneer settlement but nothing much had been done about it. Some dikes and levees had been built since the first recorded flood in 1805 but water had flown over their tops many times since then. Now, in 1913, the citizens vowed to put an end to such terrible floods as they had just experienced. Such feelings have often been held before but then, in time, forgotten. The people of Dayton, however, did not forget and they did something about it.

First, a Flood Prevention Committee was formed and it soon raised $2 million by subscription. Seven weeks after the flood an engineer named Arthur E. Morgan had been hired to show how prevention could best be accomplished. Many Daytonians wanted to get Federal help but Morgan advised that "Federal aid might prove a will-o'-the-wisp and that if the people wanted flood control they should decide to get it and pay for it, for otherwise they might repeat the history of Pittsburgh and other cities which had waited many years for help." (In that era, it should be remembered, most people took a very narrow view of the Federal government's powers and responsibilities. Activities such as flood control were considered probably unconstitutional and certainly immoral.) The people of Dayton at last decided to go along with Morgan's advice and pay for everything themselves.

Morgan's plan took in the whole basin of the Miami River. He suggested detention reservoirs on the tributaries and channel clearing on the main stream. At that time, in the United States, there were no flood control reservoirs at all. Nor any authority, public or private, with power to take on such projects.

The people of Dayton had to go to the state legislature which, only one year after the flood, passed a Conservancy Act, providing for a Conservancy District, a public corporation with power to condemn land (eminent domain) and to build works to control rivers. This act was fought in the courts for four years. Some people worried that it would be too expensive, others did not want to be moved to make room for reservoirs. Others, remembering the failure of the high dam at Johnstown, considered all such dams dangerous. Still others thought the act unconstitutional, while some did not like to see any reduction in the power of county and city governments. The District survived all attacks and $15 million in bonds were sold in the next year. Five detention reservoirs were finished by 1923 and the cost came to $30 million. The reservoirs can hold 841,000 acre-feet of water. The last of the bonds were paid off in 1950. The system has been proved to work.

The experience of Dayton in flood control was the first of its kind in the history of the United States. It was the forerunner of programs that have now run into many billions of dollars.

The Seventy-Five-Mile-
Wide River

The first tentative steps to do even a little bit about floods came along the Mississippi River. This is, of course, one of the great rivers in the world. Its statistics are awesome and so are the mighty floods it too often carries. It has inspired much song and literature; perhaps most notably some of the writing of Mark Twain.

In *Life on the Mississippi* Twain quotes from *Harper's Magazine* of February, 1863.

But the basin of the Mississippi is the Body of the Nation. Exclusive of the Lake basin and of 300,000 square miles in Texas and New Mexico, which in many respects form a part of it, this basin contains about 1,250,000 square miles. In extent it is the second great valley in the world, being exceeded only by that of the Amazon. Latitude, elevation, and rainfall all combine to render every part of the Mississippi Valley capable of supporting a dense population. As a dwelling-place for civilized man it is by far the first upon our globe.

Yet the river is prone to extremes. Twain was on board a relief boat sent out by the New Orleans *Times-Democrat* during the flood of 1882 and in his book recalls some of his observations.

This present flood of 1882 will doubtless be celebrated in the river's history for several generations before a deluge of like magnitude shall be seen. It put all the unprotected lowlands under water, from Cairo to the mouth; it broke down the levees in a great many places, on both sides of the river; and in some regions south, when the flood was at its highest, the Mississippi was *seventy-five miles* wide! A number of lives were lost, and the destruction was fearful. The crops were destroyed, houses washed away, and shelterless men and cattle forced to take refuge on scattering elevations here and there in field and forest, and wait in peril and suffering until boats put in commission by the national and local governments and by newspaper enterprise could come and rescue them. The properties of multitudes of people were under water for months, and the poorer ones must have starved if succor had not been promptly afforded.

At one point, the newspaper boat left the main stream to go up the Black River.

> Here were seen more pictures of distress. On the inside of houses the inmates had built on boxes a scaffold on which they placed the furniture. The bedposts were sawed off on top, as the ceiling was not more than four feet from the improvised floor. The buildings looked very insecure, and threatened every moment to float off.

The cattle stood breast high in the water, impassive, waiting for help. Unlike horses who would swim off in search of food, the cattle would stand until they drowned. Some people, Twain discovered, would not leave their houses until there was no more room between roof and water on which to stand. "Love for the old place was stronger than that for safety."

The city of New Orleans near the mouth of the vast river began life in 1717 and the first French settlers soon began to build levees. The settlers knew that the site of New Orleans was dangerous in terms of floods, but it was strategically so attractive that the problem of flooding seemed secondary. Soon, land grants from the King of France ordered that those who received them must construct levees. When Louisiana became a state in 1812 there were levees north up the river banks for a distance of nearly 200 miles. The protection system was all the work of individual landowners, however, not one continuous line and the construction of very uneven quality. During a huge flood in 1844, levees broke in many places and many thousands of acres lay under water.

After this failure the Federal government began to take a limited interest in the problem of floods. Congress made a small appropriation to study what might be done in the way of prevention and the Army Engineers were directed to clear the lower stretches of the river from obstructions. This was done in the name of navigation, however, because the legislators could find nothing in the Constitution about floods but power had been given explicitly to "regulate navigation".

Levees continued to be built on nothing larger than a community basis until the Civil War began. That year of 1861 two Army Engineers submitted a report to Congress pointing out that local powers could not handle the flood problem and that action would have to be on a national scale.

(Why was the United States Army involved in such matters? Because, for many years, the military academy at West Point was the only engineering school in the United States; the only engineers in the country were Army officers.)

During the war, work on the levees stopped but the river flowed on, with large floods in 1862 and 1865. Levees, too, were damaged during a number of battles along the Mississippi. Discouraged farmers abandoned a great deal of low-lying land.

Very slowly over the years Congress began to

accept the fact that it had both the responsibility and
the power to do something about the almost yearly
inundations of the Mississippi. In 1874, another commission was appointed to study the problem and even
appropriated Federal money for the relief of flood suf-
ferers. Several years later, a permanent Mississippi
River Commission was established but its powers
were contained so that no money was to be spent in
building or repairing levees if the purpose was only to
prevent the water from overflowing. Money should
not be used for anything but deepening and improv-
ing the channel.

One of the members of the commission was
Charles B. Eads who had gained an international repu-
tation with the railroad bridge he engineered across
the Mississippi at St. Louis. This immense, very high
structure still has the power to terrify train passengers
today. Eads also had considerable experience in connection with work on the levees at New Orleans. Eads
had an ingenious proposal for curbing the river by
means of jetties which would make the river everywhere of uniform width. This would stop the river
from caving in banks along the shore and the uniform
velocity would reduce friction and thus speed the discharge of floods. His idea was accepted by the Commission and work began, yet the floods continued.

Inspired by President Theodore Roosevelt, the
U.S. Congress in 1902 created the Bureau of Reclamation and also a bitter, confused situation that continues to the present day. The Bureau of Reclamation,

within the Department of the Interior, was to build dams as reservoirs and for irrigation in the arid states of the West. As a by-product, such dams would aid in flood control. The first such dam, on the Salt River in Arizona, named after Roosevelt, was hailed as a huge success.

Yet, over the years, it began to be seen that one dam can serve many purposes at the same time. It could provide flood control, irrigation, maintain water levels for navigation, create power, provide a water supply, and be used for recreation. If a dam is to do all these things simultaneously, who is to built it? The Army Engineers or the Bureau of Reclamation?

Projects have been held up for years while these two offices of the government feuded over which should have control. Recommendations for particular streams have by no means always been uniform. Both offices lobby among members of Congress, offering tempting suggestions involving large expenditures of money in the congressman's district. It has even been suggested that proposals have been made for dams that are not necessary at all. In 1902, of course, no one foresaw all this trouble. No one expected that dam building would become a business costing billions every year. Today, the bureaucracies in Reclamation and the Army Engineers are so entrenched that no permanently happy solution seems possible.

The Mississippi, historically the property of the Engineers, had major floods in 1912, 1913, and 1916, in spite of levee construction by local powers, assisted to some degree by Federal money. In 1917, Congress took a major step and voted $45 million for more levees between Rock Island, Illinois and the mouth of the river. Communities along the way would have to put up at least half the money for work of direct benefit to them and agree to maintain the structures upon completion. At the same time the Engineers were ordered to help along the Sacramento River in California, clearing out mine debris and building levees. They were also told to consider water power when considering flood control. In 1925 Congress broadened the power of the Engineers, asking them to consider all navigable waters in terms of power, flood control, irrigation, and navigation. The Engineers were to keep their hands off the Colorado River, however. This was considered to belong to Reclamation.

Events in 1927 prodded the government into taking more steps toward "promoting the general welfare." "Ol' Man River," whose powers had been celebrated by Oscar Hammerstein in *Showboat* the previous year, began an outstanding performance in the month of March. In April, rains of between 12 and 24 inches, falling over Oklahoma, Missouri, and Tennessee, drove the Mississippi to the highest levels, from Cairo on down, ever seen during its long history of floods. The death count came to three hundred and thirteen people. Six hundred thirty-seven thousand people were driven from their houses by the rampaging river. About 18 million acres of land were under water. The damage reckoned came to $300 million.

Figures such as this do not convey emotional content but enough has been said about floods so that some of the suffering can be imagined.

Governors of the six states most drastically affected asked the Federal government for help. The popular Secretary of Commerce, Herbert Hoover, known for his war relief work in Belgium, was put in charge of the Federal effort. Hoover organized local powers, the National Guard, the Army Engineers, the Coast Guard, the Navy, the Weather Bureau, and the Red Cross into a cohesive working unit.

In his autobiography, Hoover described some of the relief activity.

> For rescue work we took over some forty river steamers and attached to each of them a flotilla of small boats under the direction of Coast Guardsmen. As the motor boats we could assemble proved insufficient, the sawmills up and down the river made me 1,000 rough boats in ten days. I rented 1,000 outboard motors from the manufacturers, which we were to return.
>
> We established great towns of tents on the high ground. We built wooden platforms for the tents, laid sewers, put in electric lights, and installed huge kitchens and feeding halls. And each town had a hospital. As the flood receded we rehabilitated the people on their farms and homes, providing tents to the needy, and building material, tools, seed, animals, furniture, and what not, to start them going again. We established

sanitary measures to put down malaria, typhoid, pellagra, and generally prevention of contagious disease, all of which we continued after the flood.

As at this time we all believed in self-help, I financed the operation by three actions.

He put on a Red Cross drive on the radio and raised $15 million. The Rockefeller Foundation provided $1 million for health measures and the U.S. Chamber of Commerce provided $10 million for low cost loans. "Those were days when citizens expected to take care of one another in time of disaster and it had not occurred to them that the Federal Government should do it." Mr. Hoover's relief work no doubt contributed to his election as President the following year. His philosophy about the role of the Federal government lost him the next election, four years later.

For weeks after the flood, the railroads, highways, and bridges in the lower Mississippi Valley were virtually useless. Thousands of acres of crops were devastated and thousands of people remained penniless and homeless. Washington moved a few paces forward. It was seen that levees alone could not control the big river. Floodways were authorized to divert the flood stages and studies begun to plan reservoirs up the tributary streams. Local governments were relieved of much of their responsibility to pay for new projects. And the Army Engineers were advised that they could consider flood control measures having nothing to do with navigation.

The story of floods on the Mississippi by no means ends with the terror of 1927, but much of the more recent action to exercise control has taken place on the famous streams that flow into it.

The Politics of Dams

The Depression of the 1930s and the coming of the New Deal, with its program for bringing back prosperity by massive Federal spending, created a boom in public works and, particularly, dams. The projects were not all original with the Roosevelt Administration, however. Several had been inherited from the past.

During World War I, President Wilson, by executive order, directed the Army Engineers to build a power dam on the Tennessee River at Muscle Shoals, Alabama. This dam was to supply the power needed to operate two nitrate plants that would produce powder for ammunition. The Wilson Dam was not completed until 1925 by which time, of course, the nitrate plants were not needed but the Federal government had a perfectly good power plant on its hands. Yet Congress had not authorized the government to sell electric power.

A number of offers were made to buy or operate the dam. One of these offers, from Henry Ford, caused a great deal of indignation. None of the offers seemed good enough to Congress. It looked as if people were trying to acquire government property at next-to-nothing prices. Instead, that legislative body in 1928, and again in 1931, passed bills sponsored by Senator George W. Norris of Nebraska that would not only allow the government to sell power generated at the Muscle Shoals dam, but would authorize the government to develop power all over the depressed valley of the Tennessee River. Both these bills were vetoed by Republican Presidents.

The Tennessee River is a tributary of the Ohio River, and enters the big stream at a point 50 miles above Cairo, Illinois, where the Ohio joins the Mississippi. Yet the Tennessee is not really part of the Ohio River basin and did not, in 1933, share that basin's heavy concentration of industry and population. The region of the Tennessee Valley was lightly populated farm country, without industry or electric power. Yet it was mountainous, with streams in its many valleys, with good annual rainfall; altogether ideal country for building dams.

The Tennessee River was not notorious for its floods. Senator Norris's vision was of electric power that would make the region bloom. (Chattanooga, on the river, had suffered floods but relief of that city was not thought of as the major benefit.) President Roosevelt saw what the Norris idea might do to help relieve the Depression, and in the extraordinary year when everything he proposed would pass in Congress, the act creating the TVA was signed on May 18, 1933. The act established a Federal corporation for the purposes of "maintaining and operating the property now owned by the United States in the vicinity of Muscle Shoals, Alabama, in the interests of national defense" and of building dams on streams that "will best serve to promote navigation, effect control of destructive flood waters" and provide "for generation of electric power to avoid waste of water power."

The TVA was roundly attacked by the Republicans and just as stoutly defended. The president of one of the utility companies (actually a lawyer) which thought itself severely threatened by competition from the government led the fight against TVA in the courts. The president was Wendell Willkie and he earned so much fame in his endeavors that in 1940 he became the Republican nominee for President of the U.S.

Efforts to stamp out TVA were not successful, however. The Supreme Court finally upheld the legality of the act. In an effort to make work, dam construction was rushed and the first two, the Wheeler and Norris Dams, were completed in 1936. Four more were finished by 1940. To provide much-needed power during the first days of World War II, four more dams were rushed to completion in 1942 and four more finished by the time the war ended. Three additional major dams have been built since that time. Construction of so many high dams was possible because the Tennessee Valley, unlike the Ohio, was not heavily industrialized. Now, however, private industries have invested over $1 billion in factories and terminals along the river's banks. And now the river is navigable for 650 miles and traffic is heavy. Electric power is available there at half the average national cost. (Ironically, the appetite for electric power in the area has become so great that coal-steam generating plants have had to be built. To the shock of many, TVA is a major purchaser of coal produced by the much-hated method of strip mining.)

One effect of the creation of so many man-made lakes has been to turn the Tennessee Valley into a major recreation area. At various times, to hold back flood water or to provide water to turn generators down stream, the water level in the reservoirs may have to be raised or lowered, causing many who once criticized the TVA to complain bitterly.

In terms of flood control, these dams are estimated to have saved about $11 million a year in flood damage. For Chattanooga, by itself, savings from floods since 1936 are considered to be more than $50 million. At times when the Tennessee River is a contributor to floods, it is believed that operation of all its reservoirs lowers the flood level at Cairo, Illinois, by 2 feet. Whenever there is danger of flooding on the

During the Johnstown Flood (1889), all the debris picked up by the flood in its wild course was caught by a stone bridge. This mass of jammed material caught fire.

—The Johnstown Horror *by J. H. Walker*

It is estimated that 2,100 people lost their lives in the Johnstown Flood of 1889 due to water and fire. Every building, railroad, and bridge in the path of the flood was destroyed.

—The Johnstown Horror *by J. H. Walker*

Mother and babe cast up by the waters during the
Johnstown Flood (1889).
　　　—The Johnstown Horror *by J. H. Walker*

High water at Hickman, Kentucky, 1912.   —*U.S. Army Engineers*

An inundated street in the downtown business district of Kissimmee, Florida, during the 1947 flood.
—*U.S. Army Engineers*

A scene in Greenville, Mississippi, during the catastrophic flood of 1927. About eighteen million acres of land were underwater.

—*U.S. Army Engineers*

Immediately after Washington received news of the flood devastation, Rapid City was declared a disaster area. This is one of the many homes that was torn from its foundation by the June 1972 flood.

—*Rapid City Journal*

In June 1972, fourteen inches of rain fell on Rapid City, South Dakota, in six hours. This car was marooned by the flood.
—*Rapid City Journal*

The Rapid City Flood (1972) was the worst disaster in the history of South Dakota. Over two hundred died. Here, a flood victim is recovered by National Guard volunteers.
—*South Dakota Army National Guard*

In June 1972, in Rapid City, South Dakota, civilian volunteers, men from the Air Force base, and more than a thousand men from the National Guard continued along the path of the flood, looking for survivors and digging into the wreckage for bodies.
    —*South Dakota Army National Guard*

Governor Shapp of Pennsylvania and his wife leave the Governor's Mansion in Harrisburg in a power boat after inspecting damage caused by Hurricane Agnes (1972).

—*Wide World Photos*

High winds caused waves to lash a waterfront home in south St. Petersburg, Florida. Weather conditions reflect the approach of Hurricane Agnes (1972).

—*Wide World Photos*

Houses on Fire Island, New York, were destroyed as a result of spring storms hitting the beach.

—*NOAA*

In March 1928, the St. Francis Dam in California burst and sent 38,000 acre-feet of water rushing out. There were no survivors, and damage came to about $15 million.
—*California Department of Water Resources*

A Virginia National Guardsman keeps watch at Richmond's 14th Street Bridge after the floodwaters of the James River had receded. During the flooding caused by Hurricane Agnes (1972), water came halfway up the lampposts.
—*Wide World Photos*

Houses, a bridge, and a railroad track lie in cluttered shambles in Man, West Virginia, after a dam broke in February 1972 and sent a wall of water gushing down Buffalo Creek. More than fifty persons died.

—*Wide World Photos*

The Lower San Fernando Dam nearly inundated the valley during an earthquake that struck Los Angeles County in February 1971.

—*California Department of Water Resources*

Before the Baldwin Hills Dam in California split open early one afternoon in December 1963, the alarm was sounded and residents were evacuated from the area.

—*California Department of Water Resources*

lower Ohio or the Mississippi, the Army Engineers have the authority to instruct the TVA as to the amount of water they must hold back. The TVA has its own radio network so there is instant information about the waterlevels all over the entire system.

Although Chattanooga has benefited from the dams, it is nevertheless built on a flood plain and has had severe inundations in the past. TVA officials estimate that a flood 77 feet high is possible at Chattanooga. The dams above the city could lower the flood to 60 feet but not contain it entirely. The city itself has done little in the way of protection with flood walls and levees. Should it undergo a major flood, the newspapers will blame the dams and call them failures but TVA itself believes the citizens of Chattanooga themselves will be responsible.

The TVA is still subject to attacks. During the Eisenhower Administration something called the Dixon-Yates power contract, which would have substituted private power for that provided by the Authority, might have put it out of business. Dixon-Yates never went through, however.

TVA is criticized for putting large amounts of farm land under water. Usually 464,000 acres are indeed occupied. Three-quarters of these had suffered flooding in the past. Much of the land which may be artificially flooded is actively cultivated until a crisis arises. In answer, TVA points out the tremendous benefit it affords even to the distant Mississippi Valley. It also points out total flood control benefits along the river of more than $200 million out of a total cost of the TVA of about $1 billion. And whereas only three farms out of a hundred had electric power in 1933, practically every farm has it today.

In terms of weather, the decade of the 1930s was probably the worst in the history of the United States. Terrible droughts came to much of the land in 1930 and then again in 1931, 1933, 1934, and 1936. These helped produce the awful dust storms on the Plains States which darkened the skies even over Boston and Atlantic City. To make matters worse, dreadful floods came almost everywhere for five consecutive years.

What ultimately forced the United States into a complete commitment to the flood business began just after midnight on the New Year's Eve of 1933 when the Los Angeles suburb of La Cañada was just about obliterated by flood and landslide. Eight streams in southern California overflowed their banks that January. Extensive flooding came in nine different states during the remainder of that year. Property damage from floods in 1935 came to $127 million and 236 people were drowned as rivers rampaged. Rain in New York state broke all existing records during July. Rivers in central California did their usual winter damage at the beginning of 1936, but it was in northeastern United States that new highs were set for the amount of land being flooded at the same time. Two extraordinarily heavy rains fell within a few days of each other in March while thick blankets of snow still lay on the ground. Everywhere, from the James in Virginia to northern Maine and westward to Ohio, rivers exceeded heights never known before. Property destroyed: about $300 million. Lives lost: 107.

Newspapers loudly demanded that "something must be done" and, by June, Congress had worked out the details of its own reaction. It passed the Flood Control Act of 1936. The preamble read in part, "It is hereby recognized that destructive floods on the rivers of the United States, upsetting orderly processes and causing loss of life and property . . . constitute a menace to national welfare . . . *that flood control on navigable waters or their tributaries is a proper activity of the Federal Government* . . . that investigations and improvements of rivers and other waterways, including watersheds thereof, for flood-control purposes are in the interest of the general welfare. . . ."

The act listed 211 flood control projects to be carried out in 31 different states, but affecting nearly all states. The cost of these works would be about $300 million.

Now, among other things, it was seen that watersheds were part of the problem that the government could do something about. The Department of Agriculture was brought into the picture and told to find out what it could do to hold back runoff water and control erosion.

Passage of the Flood Control Act in 1936 did not provide instant protection everywhere, of course, and was too recent to have had any effect along the Ohio River in January, 1937. The Ohio and its many tributaries, on which much of the industry in the United States has been established, have a dangerous peculiarity. The thirteen major tributaries at the flood season may all discharge into the main stream at the same time. They always, during floods, fill the Ohio with water faster than it can discharge water at Cairo, Illinois, into the Mississippi. Because of this, every river town from Pittsburgh on down can expect floods almost every year. Many communities have built up protection against the inevitable but there is a threat every spring. In January, 1937, 30 million more acre feet of water flowed into the Ohio below Pittsburgh than flowed out during the same month at Cairo.

This spread out over 1.5 million acres of flood plains. At Cincinnati the water rose 80 feet. When the community was first settled as Fort Washington in 1773, an Indian had pointed out a high water mark of 105 feet but his cautionary was ignored.

(Edmund Swigart, a science teacher in Connecticut, says, "I can identify Indian sites anywhere in New England. They would have a nearby stream, protection from the wind, and a site relatively flat and high enough so that it would not be flooded.") People also built on low land in many other towns along the river and, as a result, 1.5 million people had to flee from their homes in January, 1937. Damage: $400 million. Deaths: 65.

In 1938, Congress handed the problem of the Ohio

River to the Army Engineers. They could not build high dams on the Ohio itself, as had been done along the Tennessee. Two million people lived there and they could not be deliberately flooded out. Flood reservoirs had to be built on the tributaries. First plans called for about 80 such reservoirs and 240 levees and flood walls. The final cost would be close to $2 billion with annual benefits of $80 million not lost through flood damage. (This would still leave an average of $20 millions of expected loss.)

As of 1970 there were nine such dams built in Pennsylvania, on the Allegheny, Mahoning, and Shenango Rivers, as well as lesser streams. In addition to flood control, certain of the dams and reservoirs provide benefits such as navigation, water quality control, a recreation area, wildlife refuges and water conservation during dry seasons.

The state of Ohio has twenty-two dams on Ohio River tributaries, fourteen of them bequeathed to the Engineers by the Muskingum River Watershed project which had been started in 1933. Two reservoirs protecting the Ohio have been built in Virginia, five in West Virginia. Kentucky has eight. Almost all of these are multipurpose projects. In time, more reservoirs will be constructed to protect the Ohio River Valley but it can never be completely safe. Measures to insure that are now two centuries too late. People and industries will continue to abide on land which, historically, the Ohio River will demand to use.

The Missouri River, longest in the United States,

was first traced from its mouth to its source 2,475 miles away, by the explorers Lewis and Clark in 1805. The historian, Francis Parkman, described the experience of the first Europeans to see the river, Joliet and Marquette. Canoeing down the Mississippi, they came upon "a torrent of yellow (that) rushed furiously athwart the calm blue current of the Mississippi, boiling and surging and sweeping in its course logs, branches and uprooted trees." They had discovered the mouth of the Missouri, "that savage river, descending from its mad career through a vast unknown of barbarism," which "poured its turbid floods into the bosom of its gentle sister."

The river had changed very little when George Fitch wrote about it for the *American Magazine* in 1907. He commented:

> There is only one river with a personality, habits, dissipations, a sense of humor, and a woman's caprice; a river that goes traveling sidewise, that interferes in politics, rearranges geography, and dabbles in real estate; a river that plays hide and seek with you today and tomorrow follows you around like a pet dog with a dynamite cracker tied to its tail.
>
> The Missouri River was located in the United States at last report. It cuts corners, runs around at night, lunches on levees, and swallows islands and small villages for dessert. Its perpetual dissatisfaction with its bed is the greatest peculiarity of the Missouri. Time after time it has gotten out of its bed in the middle of the night,

with no apparent provocation, and has hunted up a new bed, all littered with forests, cornfields, brick houses, railroad ties, and telegraph poles. Later it has suddenly taken a fancy to its old bed, which by this time has been filled with suburban architecture, and back it has gone wth a whoop and a rush as if it had really found something worthwhile.

It makes farming as fascinating as gambling. You never know whether you are going to harvest corn or catfish.

The Missouri River begins at the town of Three Forks, Montana, where the Jefferson, Monroe, and Gallatin Rivers meet after flowing down from the Bitterroot Range in the Rocky Mountains. It meanders north for almost 200 miles to Great Falls, then turns east to flow into North Dakota. After passing through Mandan, the Indian town where Lewis and Clark spent their first winter, it runs into South Dakota and down towards Nebraska where it forms the eastern part of the border between the two states. The Missouri reaches Iowa at Sioux City, generally considered the head of navigation on the river. Continuing due south it marks the Iowa-Nebraska border and then the Missouri-Kansas border. At Kansas City it meets up with the Kansas River and turns east across Missouri until it joins up with the Mississippi ten miles north of Saint Louis. All this time it has been picking up water from tributaries, even including one that begins in Canada.

During this enormous course, it passes through many different regions of climate. In a single season not only floods but droughts, cataclysmic rains, tornadoes and blizzards may be experienced along its banks. Probably no region in North America has so many different kinds of weather. And the usual problem is either excess water or far too little. During the dry periods in the 1930s the Federal government spent more than $1 billion on relief from the drought in the watershed of the Missouri River.

Attempts to contain "the Big Muddy" have been frustrating and terribly expensive. Before the United States officially became concerned with flood control, the only dams on the Missouri or its tributaries were five built by the Bureau of Reclamation. These were really for irrigation but did provide some flood benefits. The Engineers had built levees to protect Kansas City but their early efforts were generally mild. Then, during the Depression, the Public Works Administration gave the Engineers money enough to build the enormous Fort Peck Dam in Montana. At one time the largest earth-fill dam in the world, the Fort Peck structure is 4 miles wide and creates a lake 190 miles long. It was designed for flood control, power and recreation. If released, the water behind the dam would cover 19 million acres to a depth of one foot.

While World War II was still being fought, Federal dam builders were making plans for the future. William G. Sloan, an engineer in the Bureau of Reclamation, drew up a program that concerned itself primarily with irrigation, water supply, power, and flood control. At almost the same time, Colonel Lewis Pick of the Engineers outlined projects involving navigation, flood control, and power. Both reports were sent to Congress early in 1944. Before anything was voted upon, President Roosevelt proposed a Missouri Valley Authority and a program much like the TVA. Reclamation and the Engineers then got together and worked out a compromise called the Pick-Sloan Plan which divided the Missouri between them. This was passed, signed by the President, and each agency received an initial $200 million to begin work. It was thought in 1944 that the various projects would cost a bit more than $2 billion. The final cost will actually be more like $20 billion.

Now 6 major dams have been completed on the Missouri itself, all of them with many purposes. They are, in addition to Fort Peck, the Big Bend Dam in South Dakota, 2 miles long with almost 2 million acre-feet of storage capacity, Fort Randall Dam, South Dakota, with almost 6 million acre-feet, Gavins Point Dam, South Dakota with 500,000 acre-feet, huge Garrison Dam in North Dakota which holds almost 25 million acre-feet and Oahe Dam in South Dakota with 24 million. When the Sloan-Pick project is finally complete, there will be more than 100 reservoirs on tributary rivers.

Before any of the big dams, except the Fort Peck, had been completed the state of Kansas had almost two months of constant rain during the summer of 1951. Many towns were hard hit by the resulting floods but Kansas City took the worst of it. The Engineers say the damage amounted to almost $900 million but that Federal control works then in existence prevented $270 million more in losses. Critics attacked the whole Missouri River program but of course it had not really yet begun operation. For the Kansas River, which caused most of the trouble, the original plans called for 12 reservoirs but after the flood this number was increased to 22. If the whole system, now well on the way to completion, still should not harness the Missouri River, then presumably the Bureau of Reclamation and the Army Engineers will be back asking for more money to build still more dams. Since the time of the Pharoahs in Egypt, since the days of the Great Wall in China, men have enjoyed putting up great structures. American engineers also dream of bigness.

On the upper Mississippi, above St. Louis, the engineers have built a number of dams all the way to St. Paul, the head of navigation. Many residents of the area think they are for flood control but actually they are not. They do hold back some of the water, during spring thaws, which might cause trouble down stream but essentially they are for navigation. The upper Mississippi is not really a river anymore but a canal with a channel 9 feet deep at all seasons. This is adequate for heavy traffic in barges and the volume of bulk materials shipped increases every year.

From St. Louis on down, the river is always deep enough. The problem is to prevent any more happenings like the floods of 1927. After that, some people still maintained that the river could be made safe if only enough levees were built. Others believed that reservoirs on tributaries would give the answer. Thought was given to extra floodways and bypasses to carry water to the Gulf of Mexico. The Department of Agriculture believed that reforestation and more careful land use would substantially hold back flood water and they received large appropriations for these purposes. (Yet the Mississippi Valley was heavily forested when De Soto first saw it, in flood.)

Not being sure what would work best, everything is being tried today. Channels have been deepened and stabilized, more levees, floodways, reservoirs built and cutoffs established. Between the Arkansas and Red Rivers, sixteen cutoffs have been created which shorten the river by 170 miles and make it possible to keep most floods within the levees. Several hundred miles above New Orleans works have been built by which part of the Mississippi flow can be diverted into the Atchafalaya River basin, bypassing New Orleans.

In the last century, farmers along the Yazoo River, which enters the Mississippi at Vicksburg, constructed their own private levees to keep the Yazoo off their fields. This outraged the farmers downstream because there the floods, necessarily, rose even higher. Several levees were dynamited and, at last, farmers with shotguns had to patrol their levees. Now 4 reservoirs have been established on tributaries on the Yazoo and the Engineers say they protect more than 1 million acres of rich farming land.

When the Mississippi is in flood, waters on the St. Francis, Arkansas, White, Red, and Yazoo Rivers back up. Levees have been built to protect these watersheds but, in case of an extreme flood, it is planned that the levees will be deliberately breached and the areas subjected to controlled flooding.

New Orleans has not been seriously inundated by the Mississippi since the last century, but levee building has been going on so long that the city now lies many feet below the river's level at flood stage. With all the works to keep the water out, New Orleans is very much like a medieval walled city. But the city is endangered by hurricanes which churn up the water on nearby Lake Ponchartrain. During Hurricane Camille in 1970, waters from the lake spilled into New Orleans and caused millions of dollars in damage. Levees, spillways, and a system of pumps—all these do not give the city complete protection from tropical storms. More than $2 billion have been spent by the Engineers on the main stream of the Mississippi, alone, and the work continues.

As the book goes to press, the Mississippi River has chosen the spring of 1973 to produce the greatest flood in over a hundred years. Melting snows in the north and heavy local rains drove the streams up to record levels. St. Louis was up to 39 feet above flood stage and New Orleans is in similar danger. St. Louis was in particular trouble because many important new industries had been built on the flood plain since the last great inundation. And the city had not taken the steps necessary to qualify for Federal flood insurance.

New Orleans was also in danger of flooding but more in danger of losing the Mississippi River itself. Since the city is a major seaport, this would have been a catastrophe. For years the river has been trying to change its course and flow down what is now the Atchafalaya River, to the west of the city. To prevent this the Army Engineers built the Bonnet Carre spillway which could divert the river twenty-five miles above town. The spillway was opened reluctantly by the engineers, reluctantly because the small towns on the Atchafalaya were not ready for such an influx of water.

At the same time a dam on the Yazoo River in Mississippi designed to protect Eagle Lake and 200 houses, showed sign of giving under the impact of floodwaters. In spite of the hurried application by the Engineers of 4,000 tons of crushed rock, the dam did not hold, but no lives were lost. During this flood, more than seventy levees have already been topped by high water.

Nature is very hard to predict and man is undoubtedly fallible, but the Engineers may at least have produced a draw in a game which is impossible to win.

In this wild spring, storm waves damaged many summer homes on Lake Erie, and in Colorado high water caused an irrigation dam to burst, flooding the town of Kersey. With all the ingenuity of man, it would not seem that he has defeated water on the rampage.

Generations of
Inundation

The state of Maine has absolutely no Federally financed flood control projects and only Delaware can make the same claim. It is not that Maine never experiences floods. Since 1770, there have been twelve recorded on the Androscoggin River and thirteen dating from the same time on the Kennebec River. These have been caused by spring rains and thaws, by hurricanes or by local thunderstorms. It does not seem a matter of Down-East pride that would not accept outside help but that local communities had built flood works and local industries had put up dams, for power and logging, many years before there was any national control policy. These efforts were enough, even though floods were not infrequent, to keep the rivers from causing too much havoc.

RAMPAGE

Of course, measures for self-defense had also been taken at other locations in New England, long before 1936. Samuel Colt, of revolver fame, began to build a dike around his factory on the Connecticut River at Hartford in 1844. It had not been completed in 1854 and so was damaged by a flood that year but it held, afterwards, until the monster flood in the spring of 1936. At Lowell, Massachusetts, the local canal company had a large gate built and suspended at the entrance to a diversionary canal that flows now alongside the Merrimack River. This could be dropped into place, in time of need, and keep the river from flooding the canal. The citizens ridiculed the idea, not believing it would ever be needed. The gate, however, has twice been needed and used successfully.

Some communities in Massachusetts and Connecticut built modest works along river banks, to keep out floods, but, more or less unconsciously, heavy reliance was placed on the more than 100 reservoirs established in the area to operate power plants. Floods were not considered an important menace until hurricane rains in 1927 dropped 9 inches of water overnight on the Green Mountains of Vermont. The White and Winooski Rivers rapidly reached flood stage; far too rapidly for any warning to be given. In little Vermont, 84 people were killed and the staggering damages came to $40 million. Herbert Hoover then recommended three reservoirs on the Winooski and these were constructed later by the Civilian Conservation Corps.

Then came the big floods of 1936, already described, and in 1938 another hurricane flood. With a pattern that has since repeated itself, the tropical storm first struck Long Island with extremely high tides, torrential rains, and winds over 100 miles an hour. Then it swept on to Connecticut and Rhode Island with undiminished power. It was late in September and the summer resorts almost deserted; otherwise the dead would have numbered many more than the 500 who were actually counted. The storm finally moved into Massachusetts and Vermont, dashing itself out in the Green Mountains, but reserving some heavy rain for Canada. Hundreds of bridges were destroyed and much damage suffered to highways, railroads and mills built near waterways. The greatest destruction came along the ocean fronts where every man-made construction was beaten by the winds and the abnormally high waves. At one community on Fire Island, every single cottage was swept into the Great South Bay and not one beach plum left standing.

By this time it became clear that New England did need some protection and the Army Engineers drew up plans for 40 reservoirs, to store flood waters, and 20 community protection projects. The city of Hartford did not wait for this Federal program to go into effect but spent $5 million of its own money to see that the experience of 1936 was not repeated.

Weather patterns may be changing. Hurricanes had been something of a novelty in New England but they had another in 1944, then Carol in August, 1954, followed by Edna the next month, Diane in 1955, and Donna in 1960. Severe spring floods came in 1968 and made a particular mess in Massachusetts. New Englanders want protection; but there are problems.

It is difficult to find places to build reservoirs in the area. Almost every desirable location is occupied by farms or town sites. New Englanders, in particular, do not like to be told by the Federal government what will be built and where. Before any project is built, it is subjected to severe local criticism. In Vermont and New Hampshire, residents do not care much about having their farms and towns flooded to create reservoirs that will only help people in other states, downstream. Not only is the land taken from the tax rolls; a traditional way of life is being destroyed. "My great-grand-dad cleared this farm!"

In general, less is likely to be done about flood control in New England than in any other region of the United States.

Among what are known as the Mid-Atlantic states, New York lies in a very fortunate situation. Its major river, the Hudson, has sometimes overflowed in the Albany region when ice has caused jams but never in history has the Hudson flooded New York City. Presumably that is impossible since the city is so close to the sea. There have been a few floods on the Mohawk River, but the only truly endangered area is the Susquehanna River and such towns as Elmira, Corning, and Binghamton. The Delaware has caused some damage in New York, where it rises, but much more in Pennsylvania and New Jersey.

Large reservoirs, such as the Indian Lake, Sacandaga, Delta, and Hinckley, all of which control the Hudson upstate, had been built long before Federal intervention in such matters. Dams on the Delaware, Neversink, and Croton Rivers, the Scoharie and Roundout Creeks, all creating reservoirs to satisfy New York City's inordinate thirst, keep back impressive quantities of flood water each year. The Engineers have built four dams on tributaries of the Susquehanna in New York but, as will be seen, these dams are not enough to keep that river in check.

A scholarly book about dams, published several years ago, reports that, due to the reservoirs on the Delaware River holding water for New York City, there is little chance of the river causing major damage as it flows between Pennsylvania and New Jersey. Yet, a week after Hurricane Donna passed through the Delaware Valley in 1960, motorists reported seeing toilet paper caught in tree branches twenty feet above the highway near Washington Crossing.

The major plague in New Jersey seems to be the small rivers that flow ultimately into New York Bay. The worst of these is the Passaic which has caused many floods in cities such as Paterson and Newark. The Elizabeth River has also created trouble.

In Pennsylvania east of the mountains, as will be seen, the greatest menace is the Susquehanna. Many reservoirs have been built on its tributaries, much local work has been done on levees and similar constructions, but the Susquehanna continues to show that it has not been contained.

To the south, the Potomac has reached flood stage but has never really inundated Washington, D.C. Water scientists believe that the Potomac is potentially much more dangerous than it has yet shown itself to be and, as a precaution, fourteen reservoirs have been built upstream from the capital. In addition, levees have been built on the river's banks to keep water from flowing down Constitution and Pennsylvania Avenues.

In the last of the Mid-Atlantic States, Virginia, floods have not been uncommon on the James or the Rappahannock but only a few control reservoirs have been built. Damage in the past has not evidently been great enough to show that the cost of protection would provide sufficient benefits.

Between Virginia and Florida the rivers receive heavy rainfall but there are not many control works. According to a Presidential commission, "Flood control is not a pressing question in this region for the paradoxical reason that floods are so frequent, especially in the timbered bottom lands of the coastal section, as to discourage land development in the affected area." Here is one area in the United States where the residents have had the good sense to keep off the flood plains.

Flood control projects usually begin as a reaction to disaster but sometimes this can be an overreaction. Lake Okeechobee in Florida, a shallow body of water but one so large you cannot see across it, was struck in September, 1928, by hurricane winds of terrible force. The waters of the lake were pushed over the natural banks and on to the rich farm land around it. Somewhere between 1,500 and 2,000 people lost their lives.

Federal response was swift, and soon the Army Engineers began constructing high dikes and a canal through which the lake could be lowered and water drained into the Atlantic before the start of the hurricane season. Okeechobee is filled in part by the Kissimmee River, which flows in from the north. The lake, in turn, seeps out towards the south and waters the swampy Everglades. Or it did so before the Engineers walled it in behind the dikes and gave it its present very lifeless appearance.

After the construction on Okeechobee, the Everglades State Park was created and much of the swamp drained. Then, in 1965, a bad drought arrived. Conservationists demanded that water from the lake be turned into the Everglades to save the alligators, the nesting birds, and the infant shrimp that begin life there. But this could not be done! At least, not in time to save much of the wildlife. The Engineers had built such complete defenses that no channel existed between the lake and swamp. To recreate the original connection was a delicate operation that would take months. In any case. Okeechobee had already been drained in advance of the coming hurricane season and there was actually not enough water in it to help the Everglades.

Of course, the Engineers recognized their mistake and the connection between lake and swamp has been reestablished. One of the Corps's spokesmen recently wrote, "If there is one fact that is basic to all others in the water resource field, it is that *whenever you do anything to water and related land resources,* you affect *all* other use of those resources everywhere in that basin."

Out west, this can be seen clearly on the Rio Grande River as it flows through New Mexico. Because of intensive erosion ever since the region was first settled, the river has been shoaling up because of the enormous amounts of sediment settling on the river bed. The Rio Grande cannot carry the floods it used to be able to handle. The irrigated lands around it are becoming waterlogged. The ground water is now heavily mineralized. Parts of Albuquerque are below the level of the Rio Grande. A number of reservoirs have now been built on that stretch of the river to control floods, hold back sediment, and to provide water in times of drought.

The Rio Grande forms the entire boundary between Texas and Mexico, and along this stretch of the river the same problems exist. Some of the reservoirs here have been built in conjunction with Mexico. Elsewhere in the state, the Pecos River has been famous for its floods; but the dams on it have been, for the most part, designed as irrigation projects. This was originally true for most dams in Texas because the greatest problem there was drought, or the fear of it. A good flood control dam should be lowered before the season for floods begin but, in that state, floods can come in any month of the year so there is no knowing when to lower the reservoir, with its attendant loss to irrigation.

Farther west, floods on the Colorado River have already been mentioned. For many years the only controls thought possible were the levees built along the Colorado's lower reaches. Most people agreed with the verdict of Lieutenant J.C. Ives in 1857, who wrote that "the Colorado along the greater portion of its lone and majestic way shall be forever unvisited and inmolested." But Major J.W. Powell, the first man to negotiate the wildest parts of the river by boat, had visions that dams could be built in many of the remote canyons. Powell later became the first head of the Geological Survey, and he continued to talk of taming the Colorado. After the Bureau of Reclamation was established, engineers working for it began to see the possibilities. A number of plans were devised but no official action taken. Demands for positive flood control increased after the town of Yuma, which lay uneasily at the junction of the Colorado and Gila Rivers in Arizona, was flooded to a depth of four feet when levees on the Gila broke after a flash flood.

The Gila's levees had been built by the Bureau of

Reclamation, which took the lead in new surveys aimed at damming the Colorado. What finally emerged were plans for a dam so enormous that it could hold all the water the Colorado might pour into it for two entire years. It would be downstream of almost all the many tributaries prone to flash floods. This "largest dam in the world" would stretch across the river at Black Canyon, at the time of the proposal a point very difficult of access. The concrete barrier, 600 feet high, would not only check floods on the lower reaches of the Colorado, it would hold back some of the excessive flow of sediment, provide water for human consumption, irrigation, electrical power, and create a lake more than 100 miles long that could be used for water sports and fishing. If completely filled, the lake would contain enough water to flood all of New York State to a depth of one foot.

The Secretary of the Interior, whose department includes Reclamation, proposed such a dam to Congress in February, 1922, and thus he began six years of noisy controversy. Objections came from many directions. The dam would eventually be expected to pay for itself by the sale of electricity, but critics protested that there were simply not enough people in the Southwest to consume all the power that would be created. And, on principle, many people hated the idea of government being in the power business. (New power dams have since been built on the Colorado to supply the demand.) It was argued that the tremendous weight of the water, about 40 billion tons, would cause earthquakes. If the dam did not hold, what would the disaster be like downstream? (Slight tremors have been recorded from Lake Mead but nothing in the least way alarming.) Fear was expressed that the lake would fill up with sediment too rapidly and make the dam useless long before its cost could be paid off. (The lake is filling up, naturally, but present reckoning is that the dam is good for another hundred years.) Doubters also said that the river was too swift and construction would be impossible. This did not turn out to be the case.

The major outcry came from water-starved Arizona, which feared that big California would get more than its fair share of the water impounded. The struggle over who should receive how much was fought in the Congress and courts for years. Battling was by no means over when President Coolidge signed the Boulder Canyon Project Act in 1928. He had become convinced that it was technically feasible, that the power could be sold, that water could be pumped from the project to Los Angeles, and that the rich Imperial Valley would wither without a guaranteed supply of irrigation water. Yet the Act stated that flood control was the first reason for the dam.

With the arrival of the Depression in 1930, the message was to speed up construction and provide jobs. A small town was built near the site and work was continuous 24 hours a day. All sorts of records were set; for the amount of material excavated, for quantities used in building, for the number of men employed, and for speed. Construction was finished two years ahead of schedule and the lake behind the Hoover (Boulder) Dam began filling in 1936. The final cost came to $174 million and this should be repaid by 1983.

A number of other dams have since been built on the Colorado, above Hoover and below it. These are not so much for flood control, however, although they contribute to that desired end. One of these is the Parker Dam, which creates Lake Havasu, the reservoir for the Metropolitan Water District of Southern California, one of the more spectacular projects of that water-conscious state.

California would seem to be greatly gifted by nature, but there are drawbacks. One of these, of course, is earthquakes; but the frequent floods are more consistently damaging, year by year. California extends north and south over 9° of latitude; southern California is in the mid-30s latitudes, which, around much of the world, are a desert region. Although heavy rains do fall at times, the Los Angeles and San Diego regions are most often short of water. Ample rain falls in the northern part of the state, usually in the middle of winter, but it is not always wanted in such quantity. More than twenty destructive floods have been counted on the Sacramento and San Joaquin Rivers since authorities began to keep records in 1850. The quantity of water in such floods has probably not increased, but, in spite of elaborate control measures, the amount of damage certainly has. Part of the story is familiar; human encroachment on flood plains. In the old mining days in northern California this was aggravated by debris that was dumped into the streams, blocking normal flow. Now much of the water is sent into irrigation canals and these carry water where nature never intended it to go. In recent years, the ambitious, $2 billion California State Water Project, which diverts vast quantities from the northern to the southern part of the region, has changed the Sacramento River into a canal to export the precious fluid. Now the Sacramento is always a foot or two higher, throughout the year, than is natural for it. This causes seepage and erosion on the levees that began to be built beside it as long ago as 1855. In times of heavy rain, floods are easier to produce.

Dams in California are very numerous. The Bureau of Reclamation is responsible for 42 of them, the Army Engineers for 29 and the U.S. Forest Service for 15. The State of California has responsibility for more than 700 other dams, some owned by itself, some by municipalities, and some by power companies. Not all these dams came into being for flood control but in total they hold back a tremendous quantity of water. The greatest dams in the state are the Shasta Dam at the headwaters of the Sacramento, built by Reclamation for flood control, and the Oroville on the Feather River, a Sacramento tributary. The Oroville, built by the Water Project to store water that will finally be carried south, is 10 feet higher than Hoover Dam and listed as fifth largest in the world. Yet, in January,

1970, 34 inches of rain fell in the Shasta area and these two dams could not prevent the greatest floods downstream that had occurred since 1903. Without the dams, this flood would surely have been one for the history books.

For ages before southern California was settled, streams and rivers eroded the young mountains and then dropped their debris as they flowed towards the Pacific. (Los Angeles is really one vast flood plain.) At times the rivers ran dry. The river channels were not well defined. Since white settlement, the Los Angeles River has twice changed its bed. Before settlement, flood waters ran over areas now occupied by houses, factories, orange groves, railroads, highways and towns. To further complicate matters, so much of the region has been paved over that water which would normally sink into the ground now has to find some other course to follow. In addition, highway and home construction in the nearby mountains has exposed thousands of acres to swift erosion and numerous mud and landslides, which block up streams.

Beginning in 1915, Los Angeles set up a Flood Control District to handle all of these problems. Many reservoirs have been built and, as well, a number of debris dams and basins. Attempts have also been made to control erosion; but this has been hindered by forest fires, which denude areas of the surrounding mountains with gloomy regularity every autumn, just before the winter rains. Some years ago an engineer, speaking of controlling the streams around Los Angeles, said it was a task "well-nigh impossible of complete solution on the one hand and a necessity on the other."

Looking north from California, the dam builders have long been concerned with the turbulent waters of the Columbia River Basin. (Here further fuel for controversy has been added to regular dam problems by introducing the subject of fish: salmon.)

The Columbia River, which begins in Canada and has tributaries in five states, carries more water than any other river in the United States, with the exception of the Mississippi. Much of its flow comes in May and June when the winter snows in the Rocky Mountains begin to melt. These are, ordinarily, the flood months. Excessive flow occurs after a winter of unusual snow and cold, followed by a delayed spring and capped by heavy, warm spring rain. The Willamette River, which flows north from California and enters the Columbia at Portland, Oregon, often floods in the middle of winter after the not-too-infrequent warm rainstorms.

Rivalry between the Army Engineers and the Bureau of Reclamation, along with much lobbying by private power interests, has been particularly intense along the Columbia. The Engineers became dam builders originally in the interest of flood control but their first big one on the Columbia, the Bonneville, was not constructed for flood control at all but for electric power and navigation above the rapids there. Bonneville first went into use in 1937. Reclamation's enormous Grand Coulee Dam did not begin to function until 1942; a multipurpose project with flood control a major part of it. Since then both agencies have been very active on the river and its many tributaries. (The power companies have scored a few successes as well.)

The names of these dams are often quite poetic; Chief Joseph, American Falls, Hungry Horse, Flathead, Nine Mile Prairie, Priest Rapids, Owyhee, Arrowrock, Lucky Peak, Bruce's Eddy, Hells Canyon, Fern Ridge, Green Peter Lake, Ice Harbor, and Little Goose are some of the dams on the Columbia and its tributaries the Snake River, the Boise, and the Pend Oreille. Recently the United States signed a treaty with Canada for joint construction and operation of the Libby Dam on the Kootenai River, a stream that had been a major contributor to floods.

For a measure of control on the flood-prone Willamette and its tributaries, a number of reservoirs (also to be used for irrigation in the dry summer months) have been constructed and given names such as Fern Ridge, Cottage Grove, Hills Creek, Dorena, Lookout Point, Fall Creek, Cougar, Blue River, and Green Peter. Some dam builders, obviously, have an aesthetic sense.

At all dams on streams in the Columbia system, streams in which salmon have to swim en route to their spawning grounds, facilities have been built in whereby the fish can negotiate the man-made obstruction. One surly critic of this expense has said, "By the time you eat a salmon, its real cost has been $5 a pound."

The Dreaded Hurricane

Almost everyone living near the Gulf or Atlantic coasts of the United States has had some personal experience with hurricanes. Although hurricanes often bring disastrous floods, they often do nothing more traumatic than ruin a late summer vacation. "Storm warnings are posted from Block Island to Cape Hatteras." The Coast Guard advises everybody to leave the beaches and low-lying areas. Disappointed families return to their apartments in the city. For others, the situation may be much more desperate.

On the average, the warm waters of the equatorial Atlantic and the Caribbean spawn six tropical storms with winds over 75 miles an hour every year. Not all of these strike the islands of the West Indies or the Caribbean but enough of them do to cause international concern. All the lands likely to be affected regularly exchange weather news. Such communication even exists between United States and Cuba, although at present the two nations have few other relationships.

In recent years, the United States has established the National Hurricane Warning Service which involves a number of Federal agencies and some state organizations. Weather reports come from ships at sea, from commercial flights and the airplanes of the Navy and Air Force. The broadest coverage comes from the weather satellite that is in stationary orbit over the Amazon Valley on the Equator. This looks down on the same geography at all times and can send pictures of clouds and disturbances anywhere from West Africa, the Atlantic, or across the Caribbean to Central America. The pictures are received and developed at the National Hurricane Warning Center at Miami and news of any suspicious formation broadcast from there. Often, the satellite gives the first warning of any trouble that may be brewing. Once a hurricane watch has begun, U. S. planes will fly in to track it.

At the National Meteorological Center in Washington, computers, which have been loaded with all the facts known about the past behavior of hurricanes, help predict the course of the latest storm. The outputs of the machine are radioed instantly to Miami, which, in turn, sends advice to regional warning stations in San Juan, New Orleans, Suitland (Maryland),

and Boston. Predictions have not yet become an exact science, and the weather people say that they may be wrong by as much as 100 miles in deciding where the storm will strike land.

Since no one wants to be blamed for crying wolf, warning the public becomes a problem. Since there are still questions of accuracy, the Center has established this policy.

> When a hurricane approaches within 36 hours striking distance of a coastline, the portion of that coastline which has a 50–50 chance of being affected receives what is called a hurricane watch. The watch signifies that people in this area should stand by, prepared to take the necessary action to protect life and property, if a specific 'hurricane warning' becomes necessary for the area. It is the goal of the Center at Miami to provide specific hurricane warnings in time to give 12 hours of daylight to provide for the onslaught. Experience has shown that this much time is necessary to secure property and evacuate residents from exposed areas. *This is becoming more and more of a problem all the time as our residents flock to the coastline.*

The warning system has considerably reduced the loss of life from hurricanes but is still not much help in controlling wind and flood damage.

Beginning in 1961, the U. S. Navy and the Department of Commerce (which has charge of the nation's weather services) joined hands in a new project called

Project Stormfury. Their object was to seed hurricanes, to drop crystals of silver iodide into them from the air and, by cooling them, reduce the strength of the wind.

This experiement was first tried on Hurricane Esther in 1961. Scientists believed, hopefully, that they had reduced the winds of the storm by 10 percent. Again, in 1963, Hurricane Beulah was seeded; but, although the force was modified, no one was quite sure whether the changes would not have occurred naturally.

Playing with the volatile power of a hurricane can be dangerous not only to the aircraft that must fly through the rain and the window into the calm eye of the storm where the crystals are dropped, but to the coastal areas where the hurricane might strike. What if it turned out that the experiment somehow increased the storm's fury, or even if people believed it had done so? At the very minimum, the government would be besieged by lawsuits. What if a treated storm then crashed on the shores of another country and caused great damage. How would the United States handle the resulting outcry?

With all this in mind, the cloud seeders acted very cautiously and did not try again until 1969, in August, when it seemed that there was hardly any chance of Hurricane Debbie reaching populated land. On August 18 Debbie was 650 miles ENE of Puerto Rico and had winds swirling at more than 100 miles an hour. Thirteen planes were at Roosevelt Roads Naval Air

Station in Puerto Rico, waiting for just such a storm. Debbie seemed perfect, except that its distance was almost the extreme range for the aircraft, which were heavily loaded with instruments. On the first day the seeding planes dropped their loads into the eye-wall of the storm from 33,000 feet five times, with 90 minute intervals in between, After each pass, the planes observing the event found that winds were reduced, at least by 30 percent. The following day the storm was followed but no seeding was done and Debbie grew stronger again. The third day seeding planes went out again, but the reduction only came to 16 percent. It was determined later that, on three of the five passes, the seeds had been dropped in the wrong place. The fact that the storm diminished on both seeding days meant, to the experimenters, that they had made at least some effect.

Project Stormfury's limited success does not mean that we will soon be rid of hurricanes. No one as yet has an idea on how the path of a storm could be redirected. Many people point out that the rain brought to many eastern states is actually beneficial, or would be if it fell, not in torrents, but more gently. Seeding would not reduce the water content of the clouds but spread them out.

Hurricanes will continue to be with us. At almost the same hour that planes were warming up in Puerto Rico to fly far out in the Atlantic to bomb Hurricane Debbie, Hurricane Camille (in 1969) struck the Gulf Coast states of Louisiana, Mississippi, and Alabama.

Some of the storm moved north over Tennessee and Kentucky but it seemed to be diminishing. Passing over the mountains of West Virginia, it intensified and heavy rains in the middle of the night caused flash floods in many rural valleys in Virginia. One weather station reported 27 inches in 24 hours. The people of Virginia had no warning of this storm and many lost their lives. Then Camille moved on to Pennsylvania and New York, causing more floods so that these states, like the others hit, achieved the status of "disaster areas."

Camille was extremely severe, the kind of storm that comes only "once in a century." According to a presidential message to Congress, "The Weather Bureau warning for the Gulf Coast areas was timely, accurate, and adequate. As the storm approached the Gulf Coast shelter was provided for over 80,000 people by state and local authorities and the Red Cross. Evacuation operations, involving over 20,000 people in Louisiana and over 100,000 in Mississippi were well carried out." Yet there were 259 known deaths and more than 60 persons missing. "The loss of life on the Mississippi Gulf Coast was due, not to a lack of warning, but to a decision by some individuals to stay despite the warning.

"Hell, we've had hurricanes, and bad ones, or so we thought," said a Civil Defense official. "But I just couldn't conceive of 190-mile-an-hour winds."

No local authority had the power to force people to evacuate. People along the Gulf had lived through so many hurricanes in the past, they would not heed the warnings. One group of friends got together at a motel on the coast and threw a gala hurricane party. Although the sheriff gave them six different warnings, the partygoers would not leave. All these people lost their lives when the storm arrived in the middle of the night with 20-foot tides.

Pretty little towns like Pass Christian, in Mississippi, where the rich from New Orleans owned large, handsome homes along the waterfront, will probably never be the same after Camille. People have been reluctant to build expensively again in such exposed positions. Estimates of the total damage from Camille came to $1.5 billion but such figures are essentially meaningless. How can the loss be reckoned of wages not earned, merchandise not sold, taxes not paid, crops not brought in, or to resorts, such as Pass Christian, when the tourists do not come?

Agnes Arrives

During the second week in June, in the year 1972, a National Weather Service satellite picked up a tropical disturbance off Cozumel Island between Mexico and Cuba. The official hurricane season had just begun. The disturbance off Cozumel did not seem to be a hurricane, however; just one of many storms which brew up for a few days, then fade away.

No one knows just what is needed to change an ordinary storm into a great one. Something happened to this one, born near Mexico, as it moved north, picking up heat and moisture from the sea. Forecasters watching it soon realized it had changed and now deserved a name. Hurricane Agnes became the first of the year. Gathering speed, it raced across the Gulf of Mexico and ran into western Florida. There it caused some damage, such as tearing up the beach on the expensive strip of sand known as Casey Key, near Sarasota. Then it moved north into the southern states, dropping heavy rains. No longer nourished by the warm sea, it began to lose strength and the Weather Service had stopped issuing bulletins even before Agnes headed out to into the Atlantic at the Virginia Capes. The storm would soon have been forgotten but it did not behave as expected. Taking a new lease on life over the ocean, Agnes headed straight for New York City.

There had already been heavy rains in the area.

Rye, New York. June 20. Boats, cars, furniture, household appliances, all kinds of property were discovered as floodwaters receded. Blind Brook, swollen by more than 10 inches of rain in 24 hours, had gone out of control. At the Bowman Dam, just north of Rye, water began spilling over the top of the 20-foot wall. Soon basements below the dam were filled to the ceiling. The *New York Times* quoted a resident, "They say that the brook came out like the Colorado River."

New York City. June 21. The sodden metropolitan area was threatened yesterday by new downpours and flooding. In New Jersey, Civil Defense officials issued a warning. "All persons and interests in flood-prone areas are urged to remain alert during this period of flood threat and ready to take action if necessary." A weather forcaster said that "reservoirs are definitely on the high side" and advised New Jersey residents to listen to their radios and "keep the paddles going, or water wings."

Wayne, New Jersey. June 22. Civil Defense volunteers began evacuating residents along the rain-swollen Ramapo and Pompton Rivers today. "People won't leave just because the water floods their basement," one volunteer complained. "It has to get into their living rooms before they panic and then the evacuation jobs becomes dangerous, especially in the dark." The Wanaque Reservoir had been totally filled four days previously and millions of gallons of water had been spilling into the Pequannock River, a few miles above the town of Wayne. This was the 32nd flood Wayne has had in this century.

On its way north, Agnes had done some sideswiping. In Virginia hundreds of roads were closed and the Chesapeake and Ohio Railroad shut down operations. Hundreds of residents in Nelson County were evacuated by helicopter. Six people were known to have drowned and another five were missing. At Washington, D. C. the Potomac River rose five feet above flood stage. A number of motorists who tried to drive through flooded areas were drowned instead. A dam on the Occoquan River, 20 miles south of Washington, developed a large crack and volunteers tried to fill it with sand bags. In Montgomery County, Maryland, two dams began spilling over, threatening hundreds of houses. More than 100,000 homes had no electricity.

Hurricane Agnes did its worst damage during the

night of June 22. The center of the storm had moved up the Hudson River. Then it suddenly swerved west and caused "one of the worst rainstorms in decades" over northern Pennsylvania and the southern tier of counties in New York State. The result was "the worst flooding in the history of the United States." Five states were proclaimed disaster areas.

At Elmira, New York, evacuation of 20,000 residents began at 8:45 Thursday evening. The city manager said "only those who slept through the warnings or refused to leave" were stranded. At the height of the flood, Elmira's business district lay under 20 feet of water. 6,000 people had to flee from the homes in nearby Big Flats after a small earth dam collapsed. The dikes along the Chemung River at Elmira did not break down but water overflowed anyway.

At Olean, New York, on the Allegheny, the dikes were not topped but the mayor ordered about 10,000 people to evacuate. "We might as well be safe rather than sorry." At Salamanca on the Allegheny, $3 million worth of dikes had just recently been completed. The water level topped them by at least three feet. At Rochester on the Genesee River citizens were advised that the Mount Morris Dam might not be able to check a flood. The Army Engineers opened a sluice gate on the dam to prevent an overflow but the river downstream was already near flood stage and this addition made it increasingly dangerous. For 35 miles below the dam, people were urged to move to high ground. In Corning, all of the works of the famous glass company were under water and three of the town's four bridges over the Chemung River were gone. Sightseers lined the surviving bridge to gawk at the destruction. Gas, electricity, and water in Corning were not available for days after the flood. It was feared that the water had become contaminated and what could be found had to be boiled. Typhoid shots were given. National Guardsmen patrolled the streets to prevent looting. Food was difficult to acquire. The Chairman of the Board of Corning Glass, which employs 6,000 people, went on the radio to stop rumors that the company had decided to move out of town. In addition, he ordered vacation checks well in advance of schedule and offered interest-free loans of $1,000 to present and past employees who had suffered flood damage.

In New York's 's Steuben County, a total of 25 bridges were badly damaged and some of them completely washed away. The town of Wellsville could only be reached by helicopter. In Yates County, a state of emergency was declared after it appeared that a flood control sluice gate would collapse.

If Hurricane Agnes created havoc in New York State, it produced even worse conditions in Pennsylvania. The city of Wilkes-Barre had the dubious honor of being the hardest hit. Lying in the Wyoming Valley of the Susquehanna River, Wilkes-Barre had been severely damaged by two floods in March, 1936. The Susquehanna reached a level of 33 feet above normal and the damage was extreme. Dikes were subsequently built that would contain a flood 37 feet high. In 1972 the river crested at 40 feet.

On Thursday night, June 22, many people in Wilkes-Barre and Kingston, across the river, had gone to bed certain the dikes would hold. Others, however, were not so optimistic and volunteered to help save the city. A few days after it was all over, Mrs. Roz Smulowitz was interviewed by a reporter from the *New York Times.* "I had been watching the river most of the day but, frankly, I had gotten tired of hearing the old stories about the flood of '36 and knew that couldn't happen because these dikes had been built." But she became worried and, in the night, went down to the river with her husband and son to see what was going on. "There were hundreds of people in the blackness and in the rain, shouting and grunting up the slick sides of the dike. We were filling pillow cases, trash bags, anything, with dirt and sand to build a levee. Hundreds of people were working their guts out. I was separated from my son and began shouting for him. The river kept coming. We couldn't catch up and the siren blew and everyone was ordered to evacuate. It was two days of hell before we found our son." The Susquehanna had burst its confines shortly after daylight Friday morning. That day more than 100,000 people in Wilkes-Barre and neighboring communities had to flee from their homes. Fires destroyed four buildings in downtown Wilkes-Barre because firemen could not reach them.

Almost as badly hit in Pennsylvania was Harris-

burg, the state capital. Here even the governor, Milton
Shapp, was made homeless when the Susquehanna
invaded the almost-new, $2 million governor's man-
sion that had been built on the river's bank.

When it became apparent on Thursday that the
river was going to continue to rise, Governor Shapp,
his wife, and the servants carried what antiques, furniture, rugs and paintings they could to the mansion's
second floor. Then they left, inexplicably without taking their black chow, Cleo, and found refuge in the
apartment of the governor's brother-in-law in downtown Harrisburg.

Early on Friday the governor and officers of the
National Guard flew up the Susquehanna River by
helicopter, surveying the damage. They landed at
Wilkes-Barre and, after looking around, Shapp concluded that "While the devastation in Harrisburg is
disastrous, the damage appears to be even greater in
Wilkes-Barre."

As the governor was thus engaged, President
Nixon also flew over the ravished area and then came
down in Harrisburg. He talked with the state's lieutenant governor, announced that he would establish a
Federal office to help coordinate national and state
relief activities, and then flew back to his weekend
house at Camp David. On his return, the governor
telephoned the President and reported him as saying
he was shocked by what he had seen. Then Shapp and
his wife returned to their house by motorboat to rescue their dog and pick up some clothing.

Harrisburg, for a time, was nearly cut off from the rest of the state. Electricity was almost nonexistent. Police officials in the state Transportation Building directed rescue operations by candle and flashlight. Grocery stores and restaurants were closed. Police put up roadblocks to keep all traffic out of the center of the city.

Like almost everything else, businesses took a severe beating as a result of Hurricane Agnes. The destruction at Corning Glass amounted to something in the millions but the company cheerfully reported that it was fully insured against floods and business interruptions. The Erie-Lackawanna Railroad, on the other hand, blamed flood losses at Elmira and Corning as it filed bankruptcy papers. In Kingston, Pennsylvania, a cake bakery that had 400 employees announced it would close down permanently. In Corning, out of 100 small companies belonging to a local business organization, only four emerged without damage. Twenty-two hundred employees of an Ingersoll-Rand plant near Corning were laid off until the damage could be repaired some time in the fall. For some time, freight and passenger trains that normally roll across central Pennsylvania had to be routed to the west on the old New York Central tracks, via Albany and Buffalo.

It took a number of days for newspapers in the area to begin publishing again. Finally, the combined Wilkes-Barre *Times-Leader, Evening News, and Record* came out with editions full of little else but stories and pictures of the flood. These are some of the headlines:

SERVICE RESTORATION PUSHED. Utilities mount massive effort. 25,000 customers were still without service but where telephones, at least, were available, people could make three-minute outgoing emergency long-distance calls. Two emergency generators had been installed to light the streets in the central business district.

CAR LOSSES HIGH AS FLOOD WATERS SUBMERGE LOTS. David Ertley, a car dealer, said he had lost 373 vehicles, his used car building and all records. He thought the loss in cars would be at least $1.5 million even though the new cars were covered by insurance. Ertley did manage to save about 100 vehicles by driving them to Larksville and he could have managed to rescue a great many more if sightseers had not created heavy traffic jams.

106 TREATED FOR MISHAPS IN CLEAN-UP WORK. Cleaning of debris from the flood the last two days has brought more accidents and injuries than the flood itself.

FIRST MOBILE HOMES DUE FRIDAY. HUD will move 5,000 homes into Pennsylvania.

KINGSTON NURSING HOME EVACUATES 134 PATIENTS. Some 139 elderly persons were evacuated Tuesday night after the ground began to cave in at the front and rear of the building. Patients, most of them invalid, were taken to four hospitals and another nursing home. Ambulances from 20 surrounding communities made the runs. The home had suffered up to $500,000 in damages due to flooding.

CONTINUED BOILING OF WATER IS ADVISED.

AREA SITE SET UP TO EMBALM. An embalming center has been set up by the Snowdon and Disque Funeral Homes. A temporary morgue for any bodies found during clean up has been established at the Five Points Little League Center.

WARNINGS OF DANGER IGNORED BY MANY. Despite the advance warnings many persons hesitated in leaving their residences. Rescue workers noted that the majority of those who waited until high water actually forced them from their homes were elderly persons. The reluctance of these persons made massive boat-rescue operations necessary, resulting in the death of at least one rescuer by drowning.

Forty Fort Cemetery lay helpless with monuments smashed to pieces, and coffins strewn along the avenue. On Wyoming Avenue, military personnel continue the gruesome task of recovering the caskets. Draped in a white shroud, one body rested against the top of a military jeep, two soldiers holding the remains in place.

HUMOR MIXED WITH DISTRESS. An unnamed man

came into the Civil Defense medical center and took not only a typhoid shot but also a tetanus shot. With these in him, he went to a nearby bar and had some more shots. The result of all this was a "medical emergency" report. The advice was to turn the man over on his stomach, with his head to the side so he could breathe. Coffee was not recommended.

100 ANIMALS SAVED BY 2 WOMEN OF SPCA. In many instances, windows had to be broken to enter the homes and many of the animals were found in traumatic shock having been left unattended during the rising waters.

NOT MISSING, LUKESH SAYS. Nicholas Lukesh, Wyoming Borough secretary, has informed the Scranton *Tribune* that he's not "a missing person." He had heard his name listed as such over the radio. "I wasn't missing before, during or after the flood."

SCRANTON Y HAS 45 ROOMS AVAILABLE. Attempts will be made to furnish 45 more rooms which had been without furniture. The 70-year old building had been scheduled to be closed in August.

COURT OPERATIONS RESUME WEDNESDAY.

WILKES COLLEGE TO RESUME SUMMER CLASSES. The college had had an estimated $10 million in damages.

MAYOR'S MOTHER DIDN'T LIKE THAT BIG FAN. Henry Novroski, mayor of Swoyersville Borough, had just been rescued by a helicopter and was still aboard

when it swooped down to pick up a small band of
stranded people. He was astonished to see his 83-year-
old mother among them and even more astonished
when she seemed to be fighting the air crewman who
was trying to strap her into the rescue basket. Finally,
after she had been hoisted aboard, her son asked her
why she was fighting the man. Mrs. Novroski, who
has 14 children, said she was not "fighting" but "that
big fan" kept blowing her skirt up and that she had
been trying to push it down, while the crewman was
trying to strap her in.

SEWAGE TREATMENT PLANT INOPERATIVE. Sewage is
now flowing into the already polluted Susquehanna
River.

SOCIAL SECURITY AND BLACK LUNG CHECKS DELIVERY
BEING FACILITATED. (Wilkes-Barre is in the heart of the
anthracite coal country and black lung disease is all
too common.)

LOCAL BANKS MAY OPEN ON JULY 5. A major problem
at one bank was that the time dials on the vault had
been frozen by the silt in flood water and attempts
were being made to enter the vault through one of its
sides. Even with a combination, no one could open the
vault once the dials were locked in place.

DONATED FOOD IS BEING SENT INTO VALLEY. As of
Thursday, United States Department of Agriculture
donated and brought to Wyoming Valley 1,455,587
pounds of food.

PUMPING IS CONTINUING IN WILKES-BARRE. A good percentage of the water in the central business district has been drained.

REGIONAL HIGHWAY LOSS SET AT $106 MILLION. North Street Bridge replacement may take 3 years. U.S. Engineers also act on repairing 5 major dike breaks.

NEW CURFEW SET AS SIGHTSEERS HAMPER CLEANUP OPERATION. Working vehicles were hampered by thousands of automobiles clogging the streets. Many bore out-of-state licenses. The influx of cars has also created a bad dust problem The mud left has now dried and and everything in the flood area is coated with a tan layer of dirt.

HEART SEIZURE WHILE CLEANING DEBRIS FATAL. Mrs. Marjorie B. Smith, 40, suffered a heart seizure while cleaning her home.

15 BUILDINGS BLOCKING ROADS WILL BE RAZED. Property owners notified included Grace Hogan, owner of property originally at 108 Crescent Avenue, now *on* Crescent Avenue, Paul Barrett, property at 47 Diebel Avenue, property now located on Kropp Avenue, and John Bell, property formerly at 126 Crescent Avenue, now on Diebel Avenue.

TWO BREWERIES SHIP WATER IN CANS, INSTEAD OF BEER, TO LOCAL REGION. The water was airlifted out of Kennedy International Airport as part of the Wilkes-Barre Relief Drive organized by the FAA flight controller staffs at Kennedy and McArthur airports.

STATE AGENCY ISSUES RULES TO EATERIES. Flood-affected places must have inspection before re-opening,

23,531 DWELLING UNITS FLOOD-HIT IN WYOMING VALLEY. 7,237 of these were in Wilkes-Barre.

130,000 CASES OF WINE AND LIQUOR DESTROYED. Most had been submerged under flood water and were condemned by the state liquor board.

35,000 CLAIMS FILED FOR JOBLESS BENEFITS.

WOMAN IN FLOOD ZONE SURVIVED 13 DAYS ALONE. A 59-year-old woman, suffering from a heart condition, diabetes, and cataracts, rode out the flood in her third floor apartment in Wilkes-Barre. She had heard the order to evacuate but had "no place to go." She said she watched the water inch its way up to the second floor and come within eight feet of her apartment. When it began to recede she got down the steps to the second floor but could not go further due to the mud and slipperiness. She made her way back to her apartment. She had a few canned goods and some raw potatoes and green beans. Without heat or water, she grated the potatoes and mixed the green beans into the mash and ate it all raw. She existed this way for thirteen days. Sometimes when she looked out to the street someone would shout to ask if she was all right and she would answer that she was okay. Finally, two men from the telephone company knocked on her door. Hearing part of her story, they left and returned with a policeman, who took her to a hospital.

WE GOT HIT TWICE BY TROPICAL STORM. Civil Defense director Frank Townsend told a CD meeting he had learned from the Flood Forecasting Service that Hurricane Agnes had passed through the Wyoming Valley once, continued northward into New York State, then turned around and came back along the Susquehanna River again. "This may be the final explanation of why the river went to 40 feet instead of the 33-foot level originally predicted."

WILKES-BARRE WILL BE BEAUTIFUL AGAIN. This was the title of a newspaper article written by George Machinchick of King's College two weeks after the flood. Visiting parts of the city an entire month after the flood, it was difficult to believe that it ever had been beautiful. Rain was falling intermittently and the dust in the streets had turned to mud. There were piles of debris piled against the curb on streets all over the business district. Men were shovelling the muck into trucks which carted it away. This activity was going on everywhere and it was difficult to imagine how much worse it must have been at the beginning. Of the buildings facing the Public Square, only a few banks seemed to be doing any business. Pomeroy's, the major department store, was closed. There were people standing on the sidewalk outside it, waiting, it seemed, for a bus to come along. Buses did arrive but few of the people got aboard. They seemed full of lassitude, their clothing shoddy, in many ways reminiscent of the civilians in Palermo and Naples during the U.S. occupation towards the end of World War II. It was as though they had had the hell bombed out of them. They seemed broke, uncertain of the future, defeated, in Italy; and the same atmosphere permeated the people idling on the square in downtown Wilkes-Barre.

The river was only a block from the square and a look at it made one understand why the people thought it was impossible it would ever overflow as it certainly had. It lay far below the river banks, green and placid, not much more than a big stream. Military jeeps with men in green fatigues were much in evidence driving back and forth across the Market Street bridge.

Hotels in the area had not reopened a month after the event but several bars were trying to do business. They were ill-lit and the plumbing overflowed in the washrooms. No glasses were available for drinks; only paper cups and the choice of liquors was severely limited. At one bar, a workman was struggling to fix the cash register. The water had come in at least that high.

A strange smell dominated this part of the city. It was reminiscent of the odor in a house or a store after a fire has been put out, but without the component of smoke. Perhaps it is the smell of wet plaster. Wilkes-Barre, a month later, still needed drying out. City

Manager Bernie Gallagher, who assumed office right in the middle of the fight to save the dikes, said a few days later, "The central city will have to be completely rebuilt."

What Agnes Left Behind

In the aftermath of Agnes, a huge wake of stories began to appear. At Big Flats, New York, oil storage tanks cracked under the pressure of the flood and 500,000 gallons of oil and gasoline flowed through the streets. After being warned not to smoke, the 2,500 residents of Big Flat were ordered to evacuate. Near Reading, Pennsylvania, 100,000 gallons of crankcase oil were forced out of storage lagoons, coating the Schuylkill River. In York, Pennsylvania, Kenneth Heistand risked his life to go down into his basement and rescue his prized collection of old beer cans. The telephone company in Wilkes-Barre explained that many subscribers had lost service because rescue boats cut lines at the tops of many poles 25 feet high. Again, in Wilkes-Barre, the signatures of 173,000 registered voters were lost and the citizens would all have to register again.

One of the more vivid personal experiences was published a month after the flood in a most unlikely position, the sports pages of the *New York Post*. Walt Michaels is defensive coach for the football Jets and lives in Shickshinny, outside Wilkes-Barre. His house was hit, and stood, but everything in his basement was washed away; appliances, panelling, playroom, the furnace. "But that wasn't a whole lot." His mother lost her house completely and his brother not only his house, but the bar he owned and all his mementoes from pro football days.

Michaels had volunteered to work with the police rescuing people. "But they couldn't get people to leave. Right up to the end people were thinking, "it can't really come; it can't really get me."

"Some houses were just torn up right away. Damnedest thing you've ever seen. I remember looking at one house and thinking, "Gee, the water doesn't seem to be rising anymore. Then I realized it wasn't rising on the level of the house because the house was loose and floating."

"But when you're busy fighting like hell to save people and property, you don't have time to think. Afterward, when people realize that they'd lost everything, that was the terrible part."

"One guy that I went to school with went back to see what was left after the water receded. He had nothing left. Everything was gone. He just sort of slumped over and by the time they got him to a hospital he was dead. He'd had a heart attack."

"I'll never forget what I saw back there. I hope I never see it again."

The flood caused the publication of useful advice. What to do about wet plaster and mildewed furniture. For those who owned valuable books and manuscripts that had been dampened, the suggestion was to put them in cold storage until permanent restoration could be made. Garden columns described a number of complex measures that should be undertaken wherever there had been too much rain. Financial reporter Sylvia Porter devoted a whole column explaining how to handle your flood losses on your income tax return. For those who wanted to do something to help the suffering, an ad in the Philadelphia *Inquirer* suggested "spend your vacation in downtown Wilkes-Barre aiding flood victims." (As they had earlier in the year at Buffalo Creek, the Mennonite Relief Committee simply appeared in the Wyoming Valley and went to work.)

Agnes blew in at the height of the vacation season and the rainy weather caused heavy losses at resorts such as Asbury Park and Coney Island. Harness racing was greatly curtailed at Saratoga, and golf courses throughout the area had to be closed for days. In the Finger Lakes region of upstate New York swimming was banned for three weeks after the flood until the State Health Department could determine that no health hazards had developed. The Corning Glass museum, the scenic caves, and New York's wineries could receive no visitors for weeks. Warner Brother's new Jungle Habitat in New Jersey was scheduled to open just as Agnes arrived and the opening was delayed until July 15. Resorts on Chesapeake Bay such as the town of Betterton (which had been built from timber washed down the Susquehanna at the time of the Johnstown flood) found their beaches filled with debris and mud washed down from Pennsylvania.

Worse still, on Chesapeake Bay, the flood nearly destroyed the shellfish industry. Thousands of families on the Bay historically made their living harvesting clams, oysters, and crabs. The sudden influx of fresh water in vast quantities from the James, Potomac, Patuxent, and Susquehanna Rivers severely diluted the Bay. These shellfish have fairly specific needs as far as saline content goes and, if the water is too fresh, they will not eat. To add to the problems, the swollen rivers brought down raw sewage, pesticides that had been washed off farmlands, silt that covered spawning beds, and an increased flow that stirred up heavy metals.

The torrents of rain also created havoc on farms. In New Jersey, 30 percent of the lettuce crop was destroyed. Against an average production of 125 million pounds of peaches in the state, only 35 million were harvested in 1972. More than half the tomatoes and strawberries were also ruined. In New York State there was hardly any corn crop at all nor was there any hay to buy. It had to be brought in from Canada at a high price. No crops were good wherever Agnes visited. The rains leached nitrogen out of the soil and, during the prolonged periods without sun, seeds could not germinate and they rotted in the ground.

Loss of life due to Agnes does not seem astonishing. In fact, the nation might congratulate itself that there were only 118 deaths in a storm of such dimensions, one that immediately affected the lives of millions of people. From the numerous reports, it would seem there would have been even fewer casualties if warnings had not been ignored in so many cases. The warnings were criticized after the event, however. Two families in Corning, New York sued the city for $5 million, claiming that the warnings were inadequate. Members of the State Senate in New York set up a committee to investigate what was called a lack of advance warning. "The facts were minimized," one Senator said. The editor of the Wilkes-Barre *Times-Leader* also had a comment. "I heard that radio stations in New Jersey and Connecticut were predicting a flood peak of 30 to 40 feet for this valley at the same time that the Pennsylvania forecasting service was predicting only 27 feet. When I left the office at 10 P.M. I figured I'd just get a little water in the cellar." Nevertheless, the toll was low. What staggers the mind is the cost in disrupted lives, the price of it all in money gone to waste.

Funds for relief came from many private sources. The Red Cross, as usual, produced large amounts of money, which went for food, clothing, practical advice and was, in many cases, simply handed out as cash. The International Ladies Garment Workers Union, which has 90,000 members in Pennsylvania, appropriated $1 million to help them and their families. Some of this also went to assist communities in trouble.

Much of the help came from the Federal government, under the provisions of the Disaster Relief Act. At the time of Hurricane Agnes, this law directed that the Small Business Administration should provide loans up to $55,000 to homeowners for 30 years at 5 1/8 percent interest. Of this, the first $2,500 is forgiven the borrower after $500 of the loan has been paid off. Businesses may borrow up to $500,000 at the same interest. Under the same law, the Federal government is to pay the entire cost of rebuilding or repairing any public building at every level, from town upwards. With this provision, all the damage to the governor's mansion in Harrisburg, which had no insurance, was paid for out of funds provided from Washington.

One loan to a small business was given to Heidi McCarthy in Binghamton, New York. She had been billed as "The Snake Dancer" and had been working at a nightclub in Corning. The Chemung River in its rise demolished the place and, in the confusion, Mrs. McCarthy's eight-foot Indian python, Velvet, disappeared, along with Mrs. McCarthy's costumes. She soon managed to borrow another python and moved to Binghamton with her two small daughters, where she quickly found new employment. (She was a refugee from East Germany and had previously worked with a 16-foot boa constrictor in Las Vegas, but had given up this partner after it broke two of her ribs.) The Small Business people were at first startled by her loan application but soon realized that, since she had been deprived of her livelihood, she was entirely within her rights.

The demand for loans was enormous and the SBA soon found itself in danger of being without funds. President Nixon, in an election year, took the lead in asking Congress to appropriate more money for relief, ease the rules for those qualifying for loans, and even urged that private and parochial schools be assisted. All this would be retroactive and include those who had been damaged by the flooding in Rapid City, South Dakota.

There was criticism of the proposed legislation. Representative Corman of California said that after the San Fernando earthquake of 1971 it had been shown that the $2,500 forgiveness feature was an "invitation to fraud. And that is how it will be used. The little old lady whose house was washed down the river will not get help because she can't repay the loan. But somebody with a stain on his rug will get a loan because he does not have to repay."

Representative Rousselot, also of California, (where they have had ample experience of disasters) said that, after San Fernando, there were many cases in which those with serious damage got no help while whole neighborhoods of those with minor damage took advantage of the forgiveness idea to get free repairs. Frank Carlucci of the Budget office, who was sent to oversee Federal relief work in Pennsylvania, warned Congress "not to legislate now for future super disasters. If we make disaster relief too generous on a long-term basis, we remove the incentives for land-use control and insurance."

Predictably, Representative Flood, whose district included Wilkes-Barre, urged passage of the bill which called for $1.6 billion. Governor Schapp of Pennsylvania said that the bill did not go far enough. The Federal government should take over all mortgages on damaged property, with the debts to be forgiven, or at least arrange that they be repaid without interest. The bill which passed did provide the $1.6 billion, increase the forgiveness to $5,000 and reduce the interest to 1 percent. Several hundred additional million dollars were voted for temporary housing, unemployment compensation, free food coupons, to rebuild streets and for emergency work on flood control projects. Past unhappy experience with the loan program bore no weight against considerations of mercy and politics.

But generous appropriations do not provide immediate relief nor instant housing. Hearing of dissatisfaction, President Nixon sent his Vice President on a tour of the flood area. At Elmira, New York, Mr. Agnew spent two hours, dividing his time between a tour by helicopter and a press conference. In spite of all the talk, the government had only produced 100 trailers for the many thousand homeless three weeks after the event. It is difficult to see how Agnew's superficial visit would calm those impatiently waiting.

Obviously it did not; and the President, aware of news reports that spoke of "some complaints" about relief programs, then directed Secretary of Housing and Urban Development George Romney to Wilkes-Barre to put a stop to "bureaucratic haggling." Nixon acted after "Sunday news reading" and he told Romney to "take personal charge on the scene of your department's relief efforts." Nixon wanted "to achieve a more harmonious working relationship" between HUD people and local officials. Romney's trip would "give us a chance to learn from the people in the Wilkes-Barre area" what has gone wrong and make certain that "the Federal Government is producing."

Secretary Romney arrived in Wilkes-Barre on August 7 and made a somewhat more penetrating survey than Mr. Agnew had done. On August 9, Romney held a press conference which was attended by Governor Shapp and a number of flood victims. His purpose was to explain how the government was trying to speed up relief, but the governor took the occasion to repeat his demands that the U.S. take over old mortgages of those who had suffered badly.

Romney replied by charging that Shapp was being "unrealistic and demagogic" and of "fuzzing up" relief with "political issues." The press conference turned into a shouting match as the Secretary tried to be heard over the clamor of angry flood victims. One woman pressed a graphic photograph of the damage towards Romney and screamed, "You don't give a damn whether we live or die."

Romney managed to be heard, saying, "It's going to take a combination of Federal, state, local, and private efforts to resolve this situation. *The principal effort is going to have to be private.*" When Governor Shapp demanded "complete equity" in all Federal payments for flood losses, evidently referring to the great Federal generosity towards business, Romney replied, "I don't think it is helpful to inject into this a political squabble over whether people will be made whole."

During the argument, Shapp said he was diametrically opposed to Romney's views. "My philosophical belief is that when people have suffered in this way, there is no reason why the Federal government can't come in and do more than what's being done now."

It was after Romney's return from his equivocal trip that Nixon appointed Frank Carlucci of the Budget Office, a native of Wilkes-Barre to go there to "cut red tape." (About this time it was revealed that Romney's Director of Public Affairs, Jim Judge, had found jobs for two of his sons in Pennsylvania relief work.)

The situation begged to be exploited, and it was. Senator George McGovern, running as a Democrat for president, arrived in Wilkes-Barre on August 21 and toured the area by motorcade. McGovern stopped at the home of Aloysius Teufel, a computer salesman, and stood talking with the man against the sun-baked ruins of his front lawn. Teufel said later. "I don't think it's just a political play for him to come through here. He's a powerful man in the country. For Romney to come in here with his tie and his shined shoes and say 'I don't know what it's all about,' I think he made an ass of himself. For himself to say it's a political issue—there's a lot of human need down here."

At another ruined home, McGovern said to the owner, "I imagine you feel almost like you've been invaded by a foreign army."

"That's true," was the answer. "If they can spend millions of dollars in Vietnam, hundreds of thousands a day to drop bombs there, well, we are American people and I think we count. We're human beings and we count the same as any other country. And if our government can find so much to shove out to another country, why can't it find enough to put it here where it belongs?"

"Instead of bombing Vietnam, we ought to be building in this area?"

"That's right. Vietnam is a worthless cause. We have no right to be in there."

Candidate McGovern later that day met with a crowd of flood survivors who jammed the meeting hall. He heard more of the same kind of complaints and said he would try to come up with some kind of proposal by the end of the week to help them in their misery.

One thing that irritated the citizens of Wilkes-Barre was that Nixon had not visited there personally.

True, he had flown over the city but this did not seem to express sufficient concern. Acting on the advice of Frank Carlucci, President Nixon rectified this with an "impromptu" helicopter visit on September 9. He had with him a check for $4 million which he presented to Dr. Francis Michelini of Wilkes-Barre's heavily damaged Wilkes College. Then he went by motorcade to many of the same areas that George McGovern had recently visited. The President ordered his parade to stop when he noticed a man, sitting outside his heavily damaged house, drinking a can of beer. The man was Frank Vivian, a railroad employee. "I know you don't like his name mentioned," Mr. Vivian said to Mr. Nixon, "but George McGovern was through here and my wife shook his hand." Mr. Nixon then inspected the Vivian house and promised the man that he would receive a house trailer immediately.

Later in his tour, Mr. Nixon learned from Jill Barrett, of the Volunteer Service to America, that a picnic planned the following day for flood victims was imperiled because there were not enough hot dogs. The President asked, "It's a freeload, huh?" Then he inquired about how many hot dogs and soft drinks would be needed and told Mr. Carlucci "to provide anything they need to make the picnic a success." The next day six members of the White House mess crew flew in with 100 cases of soda and 2,000 hamburgers and hot dogs for the picnic which was held in a football stadium. Could any king who ruled by divine right have managed such a gesture?

We are hardly surprised when we observe human beings exploiting the going situation. It is inborn among politicians to appear, like *dei ex machina*, from the air and promise everybody anything. Such behavior produces headlines and makes the most cynical appear compassionate. Nor are we astonished when cheap and tawdry, dishonest persons take advantage of an emergency for personal gain, the people who beg for help to which they have no right. We can honestly sympathize with those, who damaged through no apparent fault of their own, demand everything that is legally theirs and, in an atmosphere when the Federal government acts as Big Brother, demand even more. But, with all this noise which disaster creates, the real victims are those who are too old and weak, or too young, powerless, and inexperienced.

Let us consider the aged. As the fall of 1972 approached, the temperature began to drop. The elderly whose houses survived continued doggedly to live in them but heating equipment was, more than anything else, the kind of convenience generally destroyed by the flooding water. Parts for oil furnaces were hard to find and repairmen even more difficult. Even in late August, 6,000 houses in the Wilkes-Barre area still had no electric power. It was nearly impossible to find plumbers to fix clogged drains or carpenters to get doors and windows to close, people to do something about mildewed walls or floors that buckled. The young and healthy might handle these problems themselves, but it was more difficult for the infirm.

Without heat, it became a general custom to go to bed when the sun went down. At least the agencies had been able to dole out extra blankets.

Allocation of the mobile homes was a source of distress. Manufacture of the units took weeks; suppliers could not keep up with the demand. Some mobile homes that had been used for relief after Hurricane Camille were trucked north and found to be very unsatisfactory. The previous occupants had sorely misused them. A general complaint: "I was promised a trailer and haven't got it." Those who did receive them were no more happy. "It's no good, that's all. No shower, no toilet, no heat, no nothing. My wife has no place to store anything in the kitchen. It's just a box car, that's all."

Emergency shelters were to have been given, first, to people with medical problems. But there was another clause in the plan. Priority should also be given to "everyone involved in the recovery effort." Naturally, local officials demanded and received the first and best. This priority was dropped, after much grumbling, but those who had the homes did not give them up. The infirm, without connections, too timid or uninformed to ask, were often the last to be helped. Many would pass a long, cold winter without any heat at all.

Football, in the valley of the Susquehanna, had been a very serious matter and, as the new school year approached, people wondered if there would be any teams or games at all. The Wyoming Valley West

High School in Kingston had seven coaches and the houses of five were flooded. Many of the students, operating on half-day schedules, were needed at home to help in rehabilitation. At the stadium of the Wyoming Valley school, water had reached halfway up the main stands and the field itself was nothing but mud.

Repairs to high-school football facilities caused criticism among people who were still without electricity or heat. But head coach, Jim Fennell, believes there would have been even more criticism if the season schedule had been canceled. One senior player felt that "football has gotten the flood off my peoples' mind." To students crowded into small, uncomfortable trailers with their families and to those expected to work long hours around the houses that survived, football practice became an excuse to get out. Coach Fennell was surprised at the numbers of candidates at the first turnout. He had thought there might be no football at all. He worried later about health problems due to the lack of heat. The whole team might be wiped out in a virus epidemic. As the season began, solidarity among the players was high. The games gave them, and the fans, the illusion of getting back to normal.

For some of the younger survivors, the flood was more serious than a mere matter of football games. As had become evident after the San Fernando earthquake of 1971, children have a hard time coping psychologically with disasters they cannot understand. Their apparently solid homes turn out to be as destructible as toys; the toys themselves have disappeared. Their parents seem lost, helpless, irritable, discouraged. Uprooted, the young wonder when they can go home again. In the town of Forty Fort, where the ancient cemetery was destroyed, children encountered decaying corpses in their own yards, in the streets, and on playgrounds. There were many nightmares. In Kingston, a little girl cries every time it rains.

Of course children must learn that life is difficult, but not all of them have to learn the fact in such a terrible way.

A Year to Remember

Although Hurricane Agnes may have caused the worst natural disaster in the history of the United States late in June, 1972, the word "disaster" was also used in many other places that summer and autumn. At almost exactly the same time as the tropical storm was sweeping across the Gulf of Mexico and heading for its American invasion, monsoon rains from the Pacific were endangering the 400,000 squatters who live in the shantytowns of Hong Kong. After three days during which 25 inches of rain fell, a hillside on Kowloon Peninsula collapsed and the landslide crushed hundreds of flimsy huts. Heavy rains or typhoons caused great damage at Hong Kong in 1966, 1968, 1970, and 1971 but always among the poor, most of them refugees from Mainland China. In 1972, however, an expensive 12-story apartment building on Victoria Peak also collapsed after a landslide. As the debris tumbled down the hill, it destroyed three other fashionable buildings. Hong Kong, of course, had been going through a frantic building boom for years with many of the new buildings being perched on dangerous locations.

New York City, which had some dislocation from Agnes, had a bit more from an unnamed tropical depression three weeks later. Breaking a 75-year old record, three inches of rain fell on a single day, July 13, tying up subway, rail, and road traffic and flooding basements in low-lying areas. The Harlem River Drive, the Bronx River Parkway, and the Belt Parkway were all under water for hours. Several New Jersey rivers reached flood stage.

The following week it was north-central Minnesota that became a disaster area. One of the heaviest rains ever recorded in the state flooded the region. At Fort Ripley, where the National Guard was in summer encampment, 11 inches of rain fell on Friday night, July 21. By morning there was six inches of water in the quarters where 9,500 men were living and four feet of water at camp headquarters. At the tourist center of Brainerd, nearby, in the heart of the lake country, thousands of vacationers were stranded, and motorists could neither get in or get out of town. Many highways were destroyed and tributaries of the Mississippi burst their banks.

Such troubles were minor, however, compared to

what was happening simultaneously in the Philippine Islands. In a newspaper dispatch from Manila, dated July 19, it was said that rain had been falling constantly for 30 hours, creating the worst floods in decades. Schools and offices in the city were closed and, in the province of Tarlac, a dam collapsed. Provincial governors were calling for helicopters and amphibious vehicles to rescue hundreds of stranded people.

Dikes out in the country began to break down the next day and an unknown number of towns began to be flooded in what was the Philippine's major agricultural district. Communications were almost impossible but President Marcos made a broadcast to what were estimated to be two million people. United States helicopters from Clark Field were sent aloft to assist and the U.S.S. *Tripoli*, a helicopter carrier, was ordered to help in rescue work.

Landslides then began, increasing the death toll, and cases of cholera began to appear as the rain kept falling, day after day. Nearly 100 cases of typhoid were also treated at the San Lazaro Hospital in Manila. Food became scarce and prices went sky-high.

Due to a poor harvest, the islands had been faced with the prospect of importing 200,000 tons of rice. With the damage done, the amount needed would be 500,000, but no country had any such surplus. The rich rice fields of Luzon, it soon was discovered, had been inundated with sand, and it might be impossible to grow rice on them again. Other crops were also destroyed, as well as much livestock, but the water buffaloes, the universal work animal, survived because they could and did swim. Just as the monsoon waters began to recede, a typhoon struck the Philippines on August 16, bringing with it more torrents of rain. The total damage was even greater than that created by all the fighting in the islands during World War II. Even before the floods, President Marcos had been faced with almost impossible social and political troubles and finally, in September, he declared martial law in hopes of bringing some order out of chaos.

But nature had not finished its attacks on the works of man for the year 1972. On August 15 a heavy, three-day rain fell on the province of Sonora in Mexico and stranded 20,000 people, wrecking 2,000 houses.

In northwest Wisconsin that same August 15, so much rain fell that it, too, became a national disaster area.

Five days later, nearly 300 people were killed in South Korea after the worst floods and landslides in the nation's history. Eighteen inches of rain fell in two days. 150,000 people were left homeless.

The United States managed to get through September, 1972, without more disasters but in October, the state of Virginia, which had had about $200 million in damage from Agnes, suffered more heavy rains and floods. The James River at Richmond was particularly threatening. Three lives were lost because people left their cars and tried to cross bridges where the streams were overflowing. On this same October 6, in Sitka,

Alaska, 11 inches of rain during the previous week, 6 inches on the previous day, caused the Blue Lake Dam to overflow, closing schools and most businesses.

The wild weather continued into November. Rains caused gigantic mud slides in the Big Sur region of California and completely wiped out the town of River Village. In Europe, on November 13th storms with torrential rains and winds strong enough to rip roofs from buildings swept across the northern countries. Twenty-five deaths were reported. Winds in the Harz Mountains of West Germany reached 125 miles an hour. The Netherlands were almost equally hard hit. Two days later a report from Paris announced that the 1972 vintage wines would be inferior, due to no sun and the heavy rains all summer long.

At the same time the United States was experiencing the wettest November on record. On the 9th, the New York metropolitan area had 5 inches of rain; traffic and power were severely disrupted. Storms that came out of the Rockies caused heavy snow in Colorado, Montana, and Wyoming and triggered tornadoes in Texas. Moving east, the storm caused severe flooding on the western shores of Lake Erie and a 25 mile strip was declared under martial law. The region was officially called a disaster area. Northern New England received more than a foot of snow. The early winter caused an acute shortage of propane gas, widely used for heating in the Midwest. During the month of December New York City had 19 days with rain and by the end of the year an all-time record had been set. The city has an average rainfall of 42 inches, but in 1972 it received 67 inches.

Why did 1972 become so particularly the year of the deluge? Some people believed it was because man has been tampering with the elements and that they were striking back. Was it pure coincidence that three times during the year record rains were falling in Asia and the United States at almost exactly the same time?

An interview with Dr. Murray Mitchell of the Environmental Data Center (Commerce Department) in Washington turned out to be most instructive. Weather is a world-wide phenomenon, and Dr. Mitchell's opinion is that the unusual events thousands of miles apart were due to the same cause. There is an overall planetary circulation of air, strongly influenced by the jet stream that races at several hundreds of miles an hour at altitudes of ten miles or more. For some reason still unknown, in 1972 worldwide whirlpools broke off from the jet stream; and it did not, as it usually does, reabsorb them. These whirlpools, closed circulation systems, simply drifted around, hardly moving at all. Storms that would usually pass rapidly over an area, instead just sat there. This was particularly true of the storms that caused floods at Rapid City and in the Philippines. In addition, what scientists call the "intertropical convergence zone" moved unusually far north in 1972 and settled on the Philippines. Such heavy rains may happen every year in the Pacific but they ordinarily fall far out at sea where no one notices them.

Dr. Mitchell's special area of research is weather variations and why they occur. He relates the extraordinary floods to the periods of drought which have a tendency to occur at intervals of one or two decades. In the United States the interval is about twenty years. Thus, there was drought in the middle 1930s, again in the middle 1950s, and the same conditions can be expected in the middle 1970s. (He says that if he lived on the high plains of the Midwest, he would be worried.) The several years prior to a drought are the wettest. Why this is so remains to be discovered.

Asked if there were more rain some years than in others, Dr. Mitchell was of the opinion that on a global basis, the same amount of rain falls every year, but *where* it falls is a much different matter.

With all this, it does not seem as if 1972 will repeat itself immediately but that, in time, it certainly will.

Stay Out of the
Flood Plain

One immediate result of Agnes was an avalanche of articles, editorials, and letters about what should be done for the future. Peter E. Black, a hydrologist in New York state, wrote,

> For too many years, engineers have sought to impress all with their skill at designing and building dams, dikes, and levees. Such awe-inspiring projects of earth and concrete sound good, provide work, and have long been hailed as one of the earmarks of civilization. But they are also suicidal. Every time a flood occurs, political machinery is put into motion to build new works "so it won't happen again." Unfortunately, the probabilities are that it will.

Black pointed out that floods are perfectly natural and that, after flood protection works are built, people consider themselves perfectly safe and so invest money in dangerous areas, thus increasing the value of the land. This, in turn, increases the demand for more protection. But these dikes, walls, and levees restrict the flow of the streams and make the water deeper during floods than it would be otherwise. When the works are toppled, the damage is multiplied. The only solution, according to Black, is to stay out of the flood plains. Flood relief should only be given to those willing to move to locations outside the plain. "Changing an entire philosophy is involved."

A *New York Times* editorial, while admitting the need for personal relief to flood victims, saw the Congressional reaction to Agnes as an opportunity for the government to move people off the flood plains by offering them subsidies, tax incentives, and other "carrots" to relocate. For those who would not move, the government would declare that it had no further responsibility in endangered areas, would provide no relief in the future, and the plains would be occupied entirely at the owner's risk. The paper deplored the idea of government-sponsored flood insurance, which merely encourages the use of flood plain areas. In a later editorial, the *Times* criticized the new low-cost loans that would encourage flood victims to rebuild on their old sites. This is "no more sensible than building a public housing project on the side of a volcano to accommodate the victims of an earlier eruption."

R. Bruce Carr of Wilkes-Barre replied to all this with some heat, if not much light. "In addition to an intense desire for simplistic solutions, the authors seem to have no knowledge of either anthropology or history." He pointed out that man has always protected his homes from every kind of danger and that, after every disaster, he rebuilds better than before.

"I lived in New York for ten years, but like so many other New Yorkers, I tired of the everyday frustrations of living there. I returned to live in my hometown of Wilkes-Barre. . . . The Wyoming Valley is a very pleasant place to live. . . . The people have character without callousness. We have had major floods in 1902, 1936, and 1972. Each time, Wilkes-Barre has returned to normal. The cleaning up and reconstruction after the recent flood is already well underway; our people are determined. At any given moment, New York City is more of a disaster area than Wilkes-Barre."

One month after the flood the Museum of Glass in Corning, New York, reopened on its former location. From Harrisburg came word that "Residents of the Wyoming Valley were reassured that they will not be forced to abandon homes that were flooded." Maurice Goddard, secretary of environmental resources, said that many residents of Wilkes-Barre and surrounding communities think they will be evacuated under the terms of a new Federal-state agreement. According to the agreement, a community may apply for Federal urban development assistance, condemn

all homes and business and pay owners pre-flood val-
ues. Goddard said that section would not be applied to
the Wyoming Valley. *"They could, but they're not going
to,"* he said. *"It doesn't make economic sense."*

The flood plain problem is not new, nor is it
restricted to the Wyoming Valley. A man and wife,
looking for land on which to build a house, usually
choose a fine day on which to drive out and take a look
around. The sun is shining as the agent shows them
an available lot. There is a lovely stand of trees along
the banks of the pretty little brook. In the valley, there
are no signs remaining of the high-water mark that
was reached perhaps ten years ago. The delighted couple eagerly hand over a down payment to the agent,
who has said nothing about floods.

During a discussion of flood plains, a wise lady
editor named Geraldine Rhoads, remarked, "Human
stupidity is also a natural resource." Seduced by the
landscape, developers have built exuberantly on the
flat plains around Albuquerque, with many fine views
of the tame Rio Grande River, and on the dry washes
made when flash floods drain down from the beautiful
mountains. Much of Utah is built on flood plains because the early settlers were in desperate need of water. Nearly every summer flash floods come out of the
Wasatch mountains, but the stubborn residents carry
on, rather than remove to safety. (Perhaps there actually is no alternative in that region.)

Flood plain zoning is much more a matter of pi-
ous talk than action. Restrictions have been enacted in

New Hampshire, Los Angeles, Milwaukee, and Miami but then weakly enforced. Suggestions to limit use of dangerous plains were offered in Cincinnati after a particularly bad experience with the Ohio River but turned down because they would cause hardship among the poor.

Among the most progressive states in this regard has been New Jersey. Some years ago, communities along the Elizabeth River began buying properties that had been devastated by floods and turning them into recreation areas and parks. New Jersey also has laws, which are enforced, regulating the construction of anything that will encroach on a stream channel. This has not prevented inundation, however. When, during Agnes, Mrs. Lillian Kosmich and her family in Wayne Township, New Jersey, had to be evacuated for the third time in a year, she said, "We bought the house 13 months ago. I specifically asked the real estate agent if this was a flood area. He assured me it wasn't. He said just the area right by the river floods. We're five blocks from the river." The Kosmichs put their house on the market but are not optimistic that anyone will buy.

Early attempts by New Jersey communities to pass flood zoning ordinances were knocked down by the state Supreme Court, which said they violated the Federal constitution because they amounted to expropriation without due compensation. In 1972 the governor of New Jersey finally presented to the State Assembly a bill that would give his office complete authority to limit or stop any further building on floodways. New Jersey is still growing, but already the most densely populated state in the Union, full of heavy industry with great political power. It will be interesting to note how effective such legislation can be.

Even where flood plains continue to be used, of course, much could be done to minimize damage. In buildings subject to immersion, electric wiring and fixtures that are waterproof should be used. Buildings that might flood away should be set on firm foundations that will not give. Sections of farms that might be inundated should be used for pasture, not crops. Millions of gallons of oil and gasoline are stored on the lowland areas of metropolitan regions; the tanks have frequently been destroyed. The combustible fluids then float downstream, causing enormous hazard. Such storage tanks should be much stronger than any are today.

Where the historic heart of a city lies on a flood plain, as in the Golden Triangle of Pittsburgh, ingenuity can greatly reduce losses. Pittsburgh's large department store, the Joseph Horne Company, now has portable aluminum panels that can be put up rapidly and withstand the pressure of 12 feet of water against the plate glass windows. The store also has pumps and extensive drainage facilities. With only a brief notice, all merchandise on the ground floor can be wheeled upstairs to safer ground. In all the new buildings in downtown Pittsburgh, the first floor is set above expected flood levels and all lower areas water-proofed. Similar precautions could be taken every-where to threatened factories and warehouses. No one will ever stop floods but their damage might be consid-erably lessened.

Many years ago the question began to be raised, "Why can't flood damage be covered by insurance?" Fire insurance, after all, is carried on every mortgaged building (bankers insist on it), fire damage in an aver-age year is almost three times as great as that from floods, but no one goes about demanding the govern-ment provide them with fire relief. Insurance pays for it all.

Some investors in the Midwest thought this way and sold flood insurance policies in St. Louis, Cairo, Illinois, and New Orleans after floods in 1895 and 1896. The company had to default on payments and go bankrupt after the Mississippi overran its banks again in 1899.

Insurance companies want nothing to do with floods. The reason lies in the nature of the business. Insurance policies are a means of sharing a burden that would be very high for individuals but tolerable if the risk can be spread among many people. For fires, accidents and thefts it is certain that some policy hold-ers will make claims but not by more than an average number that can be reckoned. Losses are a matter of chance but actuaries can determine them, to the com-pany's profit.

Floods, however, strike everyone at once. No one

who did not live in a river valley would bother to buy flood protection but everyone in a stricken valley would put in a claim after disaster struck. No company could amass enough reserve capital to pay off all the demands. In some years, flood damage would only be a few million dollars. In 1972, in the United States, it may have been on the order of $5 billion.

Recognizing that no prudent businessman would undertake the risk, Congress in 1968 passed a National Flood Insurance Act by which policies would be subsidized on residences at a cost of $40 per $10,000. The Federal government would pay 90 percent of the claim and the private insurer, who did the paperwork on the policy, would pay 10 percent. Aware that such a program, by itself, would merely encourage construction on flood prone land, this Federally backed policy would not be available for new homes. And it would only be available to communities which filed, with the Federal loan administrator, details of population, topography, and flood risk. Furthermore, these communities would have to show that they were undertaking land use programs to prevent losses by flooding.

The amount of red tape involved and the fear that the Federal government would take over local zoning meant that the program initially got off to a very slow start. (Changes in zoning laws, for someone's benefit, are not infrequently a source of corrupt profit to local politicians.) At the time of Hurricane Camille, a year after the act was passed, only two communities in the

United States qualified for this flood insurance; Fairbanks, Alaska, and Metairie, Louisiana.

Shocked by Camille, a number of communities, mostly in the Gulf Coast area, performed the necessary paper work and at a Congressional committee hearing the following summer, George Bernstein, administrator of the loan program, reported that 191 communities had become eligible. (One reason for the early lack of interest was that many people simply did not know the program existed.) After Camille, the red tape had been cut back, but land use programs were still demanded.

At this Congressional hearing Representative Robert Jones of Alabama expressed scepticism that the communities would live up to the agreements they had made to qualify for insurance. "As sure as I am a foot high, if I am still in Congress 10 years from now, they will not zone, and they will be right back up here wanting us to make further repairs. It seems to me that if we are going to make these zoning ordinances that there should be legal and binding contracts and enforceable in the courts." Then, changing the subject, he asked Mr. Bernstein why he had issued only 5,511 flood insurance policies in all the United States during the two years the policies had been available.

Mr. Bernstein replied by saying he would first like to get back to the subject of land use. "If we do not get communities to produce, this program is worthless, and it should be ended." As to the program going slowly, it had taken a long time to get communities qualified but once in the program, things went more rapidly. "In an area like Minot, North Dakota, where there was a significant flood hazard, there was something like 600 policies sold in a week. Unfortunately, there is still the second factor, which is human nature."

Mr. Jones again wondered if communities would do their part. "We just provided $9,600,000 the other day for local protective work. These things are getting out of hand."

Mr. Bernstein then added, "To speak practically, sir; I think that on a bread-and-butter basis one of our great responsibilities is going to be to withstand pressures, and I know that pressures will be placed on Congressmen, too, to water down the land use and control requirements. I think it is essential that we stick to our guns and that Congress sticks to the language of the act."

Mr. Jones replied, "It is refreshing to hear you say that, because I have had a lot of misgivings about it."

Further in his testimony, Bernstein remarked that, "The basic purpose of the act is that in the long run it will be cheaper to the public and the taxpayer to prevent flooding, to prevent losses, than to remedy after the fact with disaster relief. Prevention is still the key, not insurance."

By the middle of 1972 there were 95,000 flood insurance policy holders in the nation. The increase came in part because the program had been expanded to include apartment houses, businesses, and nonprofit institutions, rather than just residences as before. After Agnes, some 27,000 policy holders received about $100 million for damage claims, but still many people who could have had coverage did not have it. In Wilkes-Barre, which was eligible, only two policies had been sold. A resident explained, "Nobody knew it was available. There was a breakdown in communications. Someone fell asleep."

George Bernstein, the insurance administrator, who had urged state governments and insurance companies to publicize the program, said that, "Civic leaders don't give a damn about this program until there's a flood. There are only about 1,200 cities in the program and there should be 5,000."

One source blamed the lack of cooperation on greedy land speculators. They wanted to sell expensive houses in areas liable to flood, areas where the government would not sell insurance. If the purchaser knew about flood insurance, asked for it, and found he was ineligible, he might well reconsider the purchase.

Shortly after Agnes, the premium rate was reduced from $40 per $10,000 to $25 and the demand for policies dramatically increased. In Pennsylvania, the number of policies in force went up by 400 percent.

Unusual Uses of Weather

Citizens of conservative cast still, after the many years of controversy about the Federal role that began with Theodore Roosevelt, deplore the massive intervention of the United States government into almost all affairs. One such area of interference is floods, and it took more than a century before anything was done about them on a national scale. Yet floods on the great rivers, such as the Mississippi, the Columbia, and the Colorado, involve many states, and piecemeal control, as was learned in Louisiana and Mississippi, simply does not work. Weather, which is certainly an aspect of floods, is also a national matter; and it would certainly be impractical to have independent weather bureaus for each of the 50 states. There would be chaos. In spite of philosophical misgivings, it seems the Federal government is going to remain in these businesses.

(One reason, often overlooked, for the fact that states and cities continually lose power to Washington is the income tax, which was first adopted in 1913. With this device the Treasury is able to take an average of one-quarter of everyone's earnings and there simply is not enough money left around for the smaller units of government. When it comes to vast expenditures, only Washington can afford them.)

Thus, the story of floods and weather is largely a Federal one. As is the case generally in that vast bureaucracy, the organization of responsibility is very disorderly and many departments have a piece of the pie.

Consider weather, for instance. This nominally belongs to the newly named National Weather Service; but the natural disaster warning system for hurricanes and floods involves the Coast Guard, Army Engineers, the FCC, Office of Civil Defense, and the Office of Emergency Planning, among other groups. NASA is involved when it comes to weather satellites.

The field of weather modification, rainmaking, and hurricane busting, tornadoes, and control of hail, would seem to belong to the Weather Service but this is only partially true.

Rainmaking (cloud seeding) began in 1946 when Vincent Shaefer and Irving Langmuir showed that precipitation could be caused by dropping pellets of carbon dioxide from an airplane into a cloud made up of drops of water with below-freezing temperatures. Since then, it has been discovered the silver iodide crystals are even more effective for the purpose. (One ounce of silver iodide vaporized in a burner will produce 25,000,000,000,000,000,000 crystals.) Various attempts to produce rain, or snow, have been carried out in Israel, Australia, and Mexico. In the United States, modifying precipitation experiments have been carried out in California, Colorado, Arizona, Florida, Missouri, and South Dakota and backed by such groups as the National Science Foundation, the U. S. Navy, the Bureau of Reclamation, and the Weather Service itself. In a report by a committee of distinguished scientists without official government positions, issued in 1972, it was said, "For certain meteorological conditions the evidence is persuasive that it is possible to increase precipitation by substantial amounts and on other occasions to decrease precipitation by substantial amounts." Rainmaking and hurricane busting look like good bets for sometime in the future, but they are not yet everyday operational.

Nevertheless, the Defense Department has gone ahead and used weather as a weapon of war. Apparently this cloud seeding began secretly in 1966 over Laos in southeast Asia in an attempt to muddy up the Ho Chi Minh trail and interfere with the transport of war supplies. When this was finally revealed in 1972 the Defense Department reacted with customary evasiveness when questioned about it. (It transpired that the State Department had protested the use of rain-

making for war purposes even in 1967 but was over-
ruled.) Scientists learning of the Defense activities ob-
jected on moral grounds and, perhaps more impor-
tantly, because such projects interfered with
international scientific cooperation, particularly with
Russia. Worst of all, such tinkering might upset the
climate on a global scale.

At a Congressional hearing on the matter, Robert
White, head of the Weather Service, said he had no
knowledge of classified weather modification operations; which seems true but remarkable. Summoned to
testify at the hearing, the Defense Department conde-
scended to send a third-string official to answer questions. Benjamin Forman, assistant general counsel for
international affairs, refused to answer all the serious
questions about "Operation Popeye" and finally Senator Pell of Rhode Island asked him, "Are you under
instructions not to discuss weather modification in
Southeast Asia?"

"Yes, sir."

At which Pell exclaimed, "I don't think I've ever
heard such an unresponsive series of replies since I've
been on the Hill."

Ultimately, Defense did admit that it had handled
"precipitation augmentation projects" (Pentagonese is
a wonderful language) in the Philippines, in India,
Okinawa, the Midway Islands, and Texas at the re-
quest of the local governments. Okinawa and Midway, of course, are wards of Washington. Results were
good in the Philippines and Texas, but not elsewhere.

Later in 1972, a number of government officials, mostly retired and all anonymous, disclosed a number of unusual matters. The CIA had been making rain in South Vietnam as long ago as 1963. "We first used that stuff in August when the Diem regime was having all that trouble with the Buddhists. They would just stand around during demonstrations when the police threw tear gas at them, but we noticed that when the rains came they wouldn't stay on. The agency got a Beechcraft and had it rigged up with silver iodide. There was another demonstration and we seeded the area. It rained."

Another anonymous government official said, "We used to go out flying around and looking for a certain cloud formation. And we made a lot of mistakes. Once we dumped seven inches of rain in two hours on one of our Special Forces camps."

Still another official admitted that the Navy had developed a chemical that would work on warm clouds that often hovered over antiaircraft gunsites in North Vietnam. "This produced an acidic rain that would foul up mechanical equipment like radars, trucks, and tanks. This wasn't originally in our planning. It was a refinement."

While denying that it had done any of the things charged, Defense Department spokesmen remarked that even if they had, weren't rains and floods preferable to bombs?

Shortly after all this came out, there was an international ecological conference, sponsored by most of the governments in the world, that was held in Stockholm. The U. S. delegation lobbied hard against any mention of weather modification. Officially, the U.S. spends $20 million a year on research into the subject but who can guess the amount put up by the Pentagon?

In that same incredible year of 1972, the Defense Department was also accused of using floods to fight their Vietnam war. A French journalist, taken to visit bomb sites in North Vietnam, wrote that "U.S. jets went into a dive and released several bombs and rockets against the dikes on which we were standing." After denying all this from Washington, a Navy officer was permitted to say, "We're not targeting the dikes. But if a missile is a threat to you, you're certainly permitted to protect yourself. A military target is targeted and if it happens to be near a dike, then it gets hit." In any case, Defense said, North Vietnamese dikes were in a bad state of repair and, if there were floods, it would not be the bombing but the dikes which were to blame. A State Department official found an article in a Hanoi newspaper which he was glad to translate for the press. "In some places, the repair of the dike portions that were damaged by torrential rains in 1971 has not met technical requirements. A number of thin and weakened dikes which are probably full of termite colonies and holes have not been detected for repair."

The United States government, dedicated to "the pursuit of happiness" would never blow up a dike on purpose.

The Big Dam Builders

Federally, the Agriculture Department became employed in the flood business in 1936. It seemed then to many enthusiastic conservationists that floods could be stopped at the source; when the rains fell on the farms and pastures. Mottos, such as "Flood control begins at the top of the hill," "Stop the little raindrops where they fall," were produced to popularize the idea. It had suddenly become evident that unintelligent farming in the United States had caused a great deal of destruction by erosion and that deforested, denuded land could not hold the water. The Agriculture Department estimated that flood damage to farms came to over $200 million every year.

The dynamic Harold Ickes, Franklin D. Roosevelt's Secretary of the Interior, had inaugurated a soil erosion program in 1933, and this was transferred two years later to Agriculture and given the name Soil Conservation Service. This group strongly advocated contour strip-cropping and farmers generally adopted it because it kept the soil from washing away. Terraces were also built on sloping land. Contour hedges were planted. Farm ponds were encouraged and now there are more than a million of them in the United States. Such ponds not only hold back rain water but are used to water livestock, as swimming holes, as wildlife refuges, and they can be stocked with fish. In addition, they are attractive in their own right.

Tree planting had many ardent advocates in the decade of droughts and duststorms. Trees had been planted as windbreaks ever since the last century on the Midwest plains but it was believed they would also hold rainwater and even attract rain. A controlled experiment in the Tennessee Valley showed that during the summer, runoff from an observed small acreage was considerably decreased after a forest had been established but that winter floods were not affected. The idea that trees attracted rain, east of the Mississippi, turned out to be untrue when it was discovered that almost all of the rain in this area came from clouds that had originated over the Gulf of Mexico. It has not been shown that the money spent by Agriculture has materially reduced floods but certainly soil erosion is an ugly thing and nobody hates a tree.

It seems now that poor farming is not what increases flood damage but the human preemption of the land. It has been said that more than half of Los Angeles is covered with highways, parking lots and buildings. On such land, how can the rain be contained?

The increased flooding caused by urban development is not limited to Los Angeles, however. On Columbus Day in 1972, the fast-growing bedroom communities just west of Chicago suffered seven inches of rain. Cellars in hundreds of new houses filled with water, parking lots at shopping centers became lakes, and cars were swamped on the highways. This area was a stranger to floods until a few years ago.

Ground that once could absorb rain water is now covered over and it takes much less rain to cause a flood than it ever did before. In addition, builders tear up the land in the process of construction, rains come and wash the sediment into streams whose beds are then filled up. With heavy precipitation the water can do nothing but spread out. No quantity of levees or storm drains can solve this problem. It is acute around Philadelphia, Baltimore, northern New Jersey, San Diego, Phoenix, Arizona, and Salt Lake City, according to one account, and probably many other places as well. But what is the remedy? Who can stop people from wanting a house of their own or developers from exploiting this desire, no matter what the consequences? Perhaps, as the problem becomes more acute, suburban houses will be built so that the ground floor can be flooded with impunity and the business of living conducted on the second floor.

This problem of land use brings up a strange word, "reclamation." In Webster's Dictionary the synonyms are given as "regenerate, redeem, recover, and restore." How does this tie in with irrigation? As has been mentioned, the Bureau of Reclamation was established in 1902 and its first project was a dam on the Salt River in Arizona, a reservoir from which the land could be watered. But what were they "reclaiming"? Strange or not, the word was used and the lusty Bureau has gone on to build more than 100 dams since its beginnings—and they boast that none of them has ever failed.

Reclamation does not seem quite so aggressive as the Army Corps of Engineers. Reclamation is limited to states west of the Mississippi and so, in the numbers game, the Bureau has had fewer opportunities to alter the landscape. Reclamation belongs to the Department of the Interior, which has been called the "Department of Things in General." Secretaries of the Interior may come and go but the divisions under the Secretary seem to have a charmed life that is essentially beyond his control. Even Presidents find themselves often helpless when dealing with this bureaucracy, since it operates its own powerful lobby when dealing with Congress. *New York Times* columnist Russell Baker once wrote, "The common American error is to imagine the bureaucracy as a monolithic conspiracy. . . . The American bureaucracy cannot be understood as a monolith. It is rather like several hundred lidless baskets of snakes placed in a single room.

There is the confusion within each basket and there is confusion between the baskets. It is anything but a master conspiracy against the world outside the room." Nevertheless, Reclamation presents a powerful front to the public and to its rival, the Army Engineers. On one occasion, the governor of Oregon had to appeal to President Johnson to get two approved dam projects in motion. Both had been given to the Engineers, but Reclamation was somehow holding them up because it was trying to get one of them into its own grasp.

Reclamation's projects for irrigation, and the dividend of flood control that is often mentioned, have not been without their critics. The sums involved are so vast, the number of projects suggested so endless, that dam building often seems the source of the phrase "pork-barrel legislation." (One Congressman says to another, "I'll vote for a project in your district if you'll vote for a project in mine.")

In addition, the goals of irrigation themselves are sometimes questioned. Should more and more water be imported to areas that are chronically deficient in it? A member of the Montana House of Representatives, George Darrow, wrote to *Science* magazine about this. "There is no reason why the depletion of a geological deposit of groundwater should be treated any differently from that of any other renewable resource. By analogy, to provide relief and subsidies to those who continue to violate the inexorable laws of nature would be to allow the endless settlement and development of every floodplain on every river in the country, with the result that each stream could then be encased in concrete from its headwaters to its delta mouth, at public expense.

"According to the same logic, we should import gold and orebearing rocks from elsewhere so that Montana's metal mines could continue to sustain the communities that are dependent upon them. It would also follow that the United States should be obligated to reimburse the thousands of homesteaders who failed in their attempts to establish 160-acre farms on the arid Great Plains."

That empire within an empire, the Corps of Engineers, employs over 400 Army officers and about 28,-000 civilians. Many West Point graduates with the highest grades go directly into this elite organization; but it is the civilians, actually, who perpetuate its power. Although the *New York Times* said, "Dams are no longer thought, by anyone perhaps but the Army Corps of Engineers, to be effective protection against floods," the Corps proceeds dauntlessly on its way.

Statistics concerning the Engineers produce a sense of awe. Since 1936 Congress has voted about $26 billion to the Corps, which has used the money to improve navigable waters, for flood control, for power production, beach erosion control, and other work. The organization estimates that it has saved the country about $21 billion in flood damages prevented, although such a figure is at best an opinion. Engineer dams generate more than 60 billion kilowatt hours of electricity every year. These dams contain over 230 million acre-feet of water and they say that "public attendance at reservoirs" amounted, in one year, to over 280 million people, or more than the United States population.

In 1972 the Corps had 330 projects underway that were related to flood control and specifically authorized by Congress. Of these projects 120 were in the planning stage and 210 were under construction. Beyond this there were an additional 400 small flood control projects underway for which Congress had not expressly voted. The Chief of Engineers has discretionary powers to undertake small works without asking Congressional approval.

The Corps is involved on many other fronts than flood control, of course. It has a program for training foreign officials and engineers. It is involved in studies concerning a new canal between the Pacific and Atlantic. It operates recreation facilities. As another responsibility, it usually handles the Federal part in emergency cleanup after natural disasters.

The Engineers have responsibility for the United States shorelines and in a recent survey concluded that 190 miles of the shore were in a critical shrinking condition. Out of 35,000 miles of shore (this includes bays, inlets, islands, and the like), 2,700 miles are seriously eroding. One particularly endangered area is the South Shore of Long Island. Here the Corps suggested a program that would cost $200 million, to be shared half by the Corps and the other half by New York State and Suffolk County, to save the beaches. The state and county would not go along with this idea. In part they did not believe the proposed measures would work; but, more importantly, they did not want to spend tax money to help people foolish enough to build on top of the sand dunes or even, in some cases, in front of them. Hurricanes are not the only danger here. Winter storms often occur and blow for days along the Alantic coast, producing powerful currents that flow parallel to the beach and severely undercut the property. In 1962 such a storm washed away 100 houses on Fire Island.

The beaches of the United States are disappearing for several reasons. The eastern shores of Delaware, Maryland, Virginia, and the area around Atlantic City are slowly subsiding due to geological changes. (The land is sinking.) At Long Beach, California, Terminal Island had sunk 25 feet, due to the tremendous amounts of oil that had been pumped out, and the movement was only checked by the injection of fresh water underground. The beaches are also running out of sand. All the thousands of dams now in existence hold back vast quantities of silt, the products of erosion on land, that naturally reach the ocean and become sand. Tidal marshes, wetlands, are also being drained. These are not only breeding grounds for fish and feeding grounds for birds, they often take the brunt of storm waves and lessen their destructive force inland. There are numerous private movements afoot to save these wetlands from developers, and the Department of the Interior has interested itself in the problem. In California the beaches are just naturally disappearing, due to the lack of sand, but also due to the natural forces of tides, wind and surf. Much of the endangered property is in private hands, and the Engineers can do nothing to help the owners. Sometimes, however, the beaches are disrupted by actions of the United States Government. The entrance to the U.S. Naval Weapons Station at Anaheim Bay had been altered by much naval construction. This caused considerable damage to nearby Surfside and Sunset beaches. Since this was the fault of the government, the Engineers undertook to pay 67% of the cost of pouring 2 million cubic yards of sand on these beaches every six years. Engineers also sometimes pump sand from Great South Bay, over the dunes and to the beaches of Fire Island, but in this case they are reimbursed by Suffolk County.

As long ago as 1899, Congress had authorized the Engineers to control dumping of garbage and wastes in the ocean and in navigable waters but the Engineers seem to have done hardly anything with their authority and it was mostly forgotten. In recent years, however, pressure for an improvement in the environment has led to changes in approach. The powers of the Corps had been narrowly interpreted to mean only dumping that would interfere with military or commercial navigation. Now it would be viewed in terms of the public interest which include "Navigation, fish and wildlife, water quality, economics, conservation, aesthetics, recreation, water supply, flood damage prevention, ecosystems and, in general, the needs and welfare of the people."

Now the Corps tries to see that none of its projects damage the natural environment. Before the Libby Dam in northern Montana was even begun, an inventory of all fish and wildlife in the area was made and while construction was underway, care taken that water quality was preserved above and below the dam. In the Cape Fear River area of North Carolina, dredged materials have been placed to create islands which wildfowl now use as rookeries. In some North Carolina sounds, the Corps has also created marshlands. Along the "Mississippi Flyway" they have asked Congress for funds to purchase wildlife refuges in Louisiana and Arkansas. Prairie dogs are now protected near Corps dams on the Missouri.

The Engineers spent $12 million to prevent the catastrophe that would have occurred to the spawning grounds of steelhead trout by the construction of the Dworshak Dam in northeastern Idaho. Like the salmon, these trout are born high in the rocky, mountain streams. They spend two years maturing in fresh water and then swim downstream to spend between one and five years in the Pacific. When it is time to breed, they return to their birthplace; but the Dworshak Dam would make it impossible to get there. Now the steelheads are caught as they approach the dam.

They are taken to a hatchery on the North Fork Clearwater where the eggs are removed from the females and fertilized by the males. The small fry are tenderly cared for until they are ready to go to sea. The pathway up the river to the place where the young are artificially bred is not obstructed and, when this new generation returns to breed, it will seek the place where it was born and man's intervention will no longer be needed.

When the Engineers became involved with floods, the enabling legislation made it clear that each of their projects had to show a positive benefit-cost ratio. For each tax dollar spent, there must be a return of at least $1.10 in flood control, power, irrigation, water or some other value. This was to be a protection against unnecessary and unneeded public works. At the hearings after the Rapid City, South Dakota, flood, congressmen heard that the Engineers had recommended a series of eight dams in the Black Hills to prevent just what happened, but they could not demonstrate a favorable benefit-cost ratio. According to Congressman Frank Clark of Pennsylvania, who led the meeting, the Engineers get their directives from the Budget Bureau and thus are notoriously conservative. "Many recommended projects by the Corps of Engineers have been lost at this level." All the Congressmen present agreed that the preservation of life is a benefit but how can the Budget people reckon the value of such an intangible?

Critics of the Engineers do not agree that it is all

that conservative when calculating the cost-benefit ratio. A retired member of the Corps, Edwin R. Decker, was quoted as saying, "It's nothing but a ritual. They come down the aisle swinging their incense and chanting 'benefit-cost'. You can adjust the b-c ratio to justify any project. I did it myself a few times."

How can benefits be calculated? A project on the Tombigbee River that runs in Alabama and Mississippi first had a ratio of only $1.01 for every $1 spent. This went back to the drawing boards and returned as $1.60 per $1.00.

What is the cash value of recreation? Often a claim is made that a project will be a saving for shippers. Is it better to ship by barge or railroad? In the estimates of flood savings, what about the value of saving buildings that would never have been constructed if there were no dam? The benefits are reckoned in terms of fifty years; how can this be realistic?

Criticism of the Corps is not new. President Chester A. Arthur in 1882 vetoed one of their appropriation bills with the comment that it included something for everybody. Every Congressman received a project for his district. "Thus, as the bill becomes more objectionable, it secures more support."

Justice William Douglas called the Corps "public enemy number one." The Engineers have been named the "pork barrel soldiers" and the boondoggle corps." Arthur Morgan, engineer for the Dayton, Ohio, flood control project who later became chairman of the T.V.A., wrote a book called *Dams and Other Disasters.* In it, he criticized the Engineers' power to initiate projects under $1 million without any control from Congress. "If a member of Congress is not on good terms with the Corps, it is very easy to find that 'with the calendar of work so full, we may not get to this for a few years.' But if it is time to do a Congressman a favor, there may be a reexamination of the appraisal of costs and benefits which will make the project seem 'feasible'."

Pushing and pulling seems endemic to the situation. After Agnes, Governor Rockefeller of New York wrote a letter urging the Engineers to press a number of flood-control projects that seemed slowed down. A plan for a program on the Saw Mill River basin in New York, that had been before a House committee for 14 years, was the subject of an angry meeting in the town of Yonkers. A representative of the Corps spoke to the citizens and told them of more planning that was being undertaken. This irritated the group, which had sustained $2 million in damage the previous month, even further. Those present blamed the new Hawthorne Traffic Circle and new highway and residential construction in the northern part of Westchester County for making the problem. "This is not Yonkers water and this is not Yonkers drainage that is causing the extreme damage." Representative Peter Peyser, from the county, assured the people present that he would assert pressure when he returned to Washington to relieve the problems caused by the Saw Mill River.

In Illinois in 1972 the Engineers were strongly attacked during a symposium on Societal Problems of Water Resources. John C. Marlin of the University of Illinois deplored the many new dams and barge canals planned for the state. He did not like the way the Corps of Engineers promoted dam projects and ignored opposition to them. He doubted the value of many new dams planned, in view of their impact on the people and the environment, and mentioned several that were not only unnecessary but actually harmful.

Two other members of the University of Illinois faculty talked specifically about the Allerton Park-Oakley Dam project, saying that nitrates would quickly fill the lake with weeds, that flooding problems on the Sangamon River would actually be increased, and the wildlife habitat reduced.

In 1972 the Engineers ran into considerable trouble over something called the Tocks Island Dam, just above the Delaware Water Gap, on the Delaware River between New Jersey and Pennsylvania. This had been proposed ten years before as a $90 million program for flood control and water storage. The dam would have been 150 feet high and created a lake 35 miles long, flooding 12,000 acres of lovely, rural farmland. As time went by, the government purchased $69 million worth of land and dispossessed the owners.

Meanwhile cost estimates rose to $369 million. Opponents called it the "$369 million mistake," a major river-wrecking project.

Governor William Cahill of New Jersey, who had sponsored the idea while he was a member of Congress, later began to have reservations. The engineers had dreamed that the dam would create a recreation area that would draw 10 million people a year and the Governor worried about the cost of new highways needed to handle such an invasion. He began to worry about how much flood protection the dam would actually provide, the need for elaborate sewage disposal facilities, and the impact of so many tourists on the region. He temporarily withdrew his support and waited for a Federal government offer to pay 90 percent of the cost of the new roads, which would come to $683 million, and the sewage facilities, which would demand another $100 million. With a blow on the side of the enviromentalists, Cahill said, "The arrival of large numbers of visitors would inevitably lead to intrusive and tawdry land use." Neverthless, he urged that the land bought by the Federal government be turned over for use as a national park.

The Army Corps of Engineers had been set back, temporarily. However, they are more than 100 years old, while governors merely come and go.

What Comes Next?

The experiences of 1972 surely indicated that mankind had not yet managed to control nature. 1972 may not have been entirely in vain, however.

As long ago as 1966, a group which called itself the United States Committee on Large Dams, a consortium of electric power companies, the Bureau of Reclamation, the Army Engineers, and the T.V.A., had compiled a booklet called *Supervision of Dams by State Authorities* and it was pertinent and very much in print six years later.

This publication revealed that five states required no license or permit before a dam could be constructed. Thirteen states had no printed regulations on the subject. Seventeen states reported that they had no authority over the way dams were operated, and ten states had no regulations saying that dams must be supervised by a professional engineer. Many states did not even have a record or list of dams in their jurisdiction. In Oregon and Massachusetts, it turned out, if you, as a citizen, complained about the safety of a dam and, after inspection, it turned out to be no hazard, you would yourself have to pay for the costs of the inspection. The state of Georgia did not even bother to answer the committee's questionnaire. South Dakota would not say whether it exercised any control over design, construction, operation, or maintenance of dams. Nor did it say how many dams there were in the state. South Carolina seemed only interested in dams as they related to mosquito control. West Virginia had no rules for operation or maintenance of dams. Nineteen states reported that their present regulations concerning dams should be reconsidered but West Virginia and South Dakota both felt that everything was satisfactory.

When asked for comment, Alabama remarked, "Already there has been one death and thousands of dollars in property damage from failure of medium-sized dams in the state. These failures appear to have been the result of improper design which was not done by licensed engineers." Mississippi said, "A recommendation of suitable regulations would be appreciated." Pennsylvania said, "After the structure has been constructed, it is almost too late to consider what should have been done during the initial construction phase. Our records indicate that remedial measures accomplished after completion of a dam have been proven to be generally unsatisfactory." A look at the complete survey suggests that most states have not been terribly alert to the dangers presented by dams in their territory.

Not so the state of California, however. Since the St. Francis dam broke in 1928, they have had the most rigid inspection laws in the country. (In the last few years, copies of this law have been adopted by a number of other states and Australia, Venezuela, and Argentina, as well.) Answering the questionnaire, California said that it inspected 1,061 dams at least once a year and more often if it seemed necessary. It did not bother with Federally owned dams which, presumably, were well tended. When asked how they enforced their orders concerning dam safety, California replied, "By persuasion, letter order, formal order and civil court action. Criminal action is also provided." California has a special Division of Safety of Dams, with a staff of 37 engineers and a budget of about $1,250,000 a year. It does not provide any printed formula for dams because such codes would not be uniformly applicable to widely varying conditions and "such a code could inhibit advancement in the science of dam design and construction." According to the magazine *Water Power*, "California Leads the Way in Providing Safe Dams" and yet the Van Norman dams in San Fernando nearly failed in 1971. Such an earthquake had never been metered before but, as has been mentioned, C. J. Cortright, head of the dam safety program, has said that all structures will be looked at again with the new knowledge of earthquake potentials.

Several years ago the Bureau of Reclamation became particularly concerned about the safety of their dams. Unprecedented floods had threatened some of them. "Because the Bureau is one of the world's leading designers and builders of dams, its engineers consider safety of the more than 200 storage dams on Reclamation projects a paramount responsibility." The Bureau in the past seventy years had acquired a number of dams, which it had not built itself, in order to coordinate projects on various streams, and some of these had not been constructed to its "exacting standards." In addition, more information about possible maximum floods had been collected, areas below many dams had become densely populated, and many of the older dams did not meet present-day standards.

In the first year of the new Reclamation program, it was discovered that, out of 15 dams studied, 12 needed major remedial measures. At one, the concrete spillway had deteriorated and it rested on very erodible soil. At another, the spillway was just too small for floods that could be expected. At an earth dam, built in 1927, which had held so far, recent floods had broken all records and would probably continue to do so.

In addition to such defects, Reclamation found that its communications with some dams in remote areas was inadequate. During a flood in Montana in 1964, it turned out that several caretakers of dams could not be reached because the telephone lines had been destroyed and roads washed out. The Bureau decided to acquire a few helicopters. After the landslide in Italy which caused the Valont Dam to overflow, Reclamation geologists began to examine the sites at all their reservoirs with this possibility in mind. It transpired that a number of them could be endangered in this way and now a periodic watch is kept, particularly during times of heavy rain.

During this survey, it was discovered that dams, not owned by the Bureau, but upstream from their structures, were often inadequate and might give under stress. Sometimes a dam built only to a height of 10 feet might have been enlarged to a height of 50 feet with hardly any thought given to the engineering consequences. The Bureau began to enlist the help of state governments to correct the deficiencies of these dams which periled its own. It works very hard to make sure that its record of safety remains unbroken.

Of course, the Army Corps of Engineers have not been less alert and were conducting their own investigations about safety, which included strip mining, even before the disasters of 1972.

Then Congress, in its wisdom, passed the bill, H. R. 15951, and threw all the dam builders into a tizzy. The bill was sponsored by Representative John Blatnick of Minnesota and cosponsored by his fellow members on the House Committee on Public Works. It was quietly introduced at a private committee hearing and sent to the floor for a vote, without a public hearing. It ordered that the Army Corps of Engineers inspect every dam in the country, no matter who owned it, for safety and report to Congress within two years on the results and with safety recommendations.

At the same time, a Senate Committee on Interior and Insular Affairs was holding hearings on a somewhat similar measure and receiving adverse testimony from the Army Engineers, the Bureau of Reclamation, and the Soil Conservation Service. They had not had time to think about the idea, and the administration was against it. In spite of the 1966 study which showed how woefully the states were handling dam safety, an assistant Director of the Budget said, "The existing mechanisms, under the states' leadership, should meet the problem."

It seemed at the time as if the bill was something of a power play between a House and a Senate committee and that the House people decided to rush their measure through rather than have some watered-down version presented. (It was inspired, of course, by what had happened at Buffalo Creek and Rapid City.) The House measure passed quickly and went to President Nixon who reluctantly signed it. He called the act "most unfortunate" because it departed from the "sound principle that the safety of nonfederal dams should primarily rest with the states."

Again, the old conflict that has been going on for at least seventy years. Although it had been demonstrated that the states had been doing little or nothing about dam safety, although dams had burst recently in two states that were completely lackadaisical about the subject, let the states handle it, let the people drown, and we will appropriate a billion dollars as relief for the people's misfortune.

Beyond the drive for greater dam safety, what had been gained from the disasters of 1972? One sorely beset state, New Jersey, passed a law which permits the government to limit, in any way it sees fit, the development and use of lands subject to floods. This is so simple and enlightened that it might be used as a model for other states.

In Rapid City, South Dakota, the Mayor had declared he would fight to keep housing from being rebuilt in the area susceptible to floods from Rapid Creek. The cry "Don't build on flood plains" has been heard, and many people who were unaware of the problem may have learned to be cautious.

In December a National Water Commission issued a report noting that in spite of all the billions spent by the Federal Government, the total loss from floods continues to grow. It criticized the dam building frenzy of the Army Engineers. The report said that the emphasis on flood control should be in discouraging building on flood plains and protecting people now living there. It should also "eliminate the unconscionable windfall gains accruing to some landowners when protection provided at no expense to them results in large increases in the value of their lands." Instead of protecting people from floods, many flood control projects merely open up unsafe areas for further development. If the year produced nothing else, if the Engineers are forced to reconsider some of their discredited projects, then the year would not have been entirely in vain.

We will always have floods, but with better warning systems (the National Weather Service is hard at work on this), wiser use of flood plains, citizens hopefully more wary and willing to leave their threatened homes (many lives were lost because of people's sheer stubbornness) disasters can be minimized. And, possibly, the Federal government has learned a few lessons in handling disaster relief. The victims need much more efficient aid and far fewer self-serving press releases. The scandals at Wilkes-Barre, for instance, are just now beginning to be revealed.

Index

INDEX

INDEX